INCOME FROM OUTCOMES

Tap Into Unrealised Income Through a Customer Outcome Program

PAUL HENDERSON

INCOME FROM OUTCOMES
Tap Into Unrealised Income Through a Customer Outcome Program

Paul Henderson
For more information, address: pjh@outcomeleaders.com

Printed Worldwide
First Printing 2022
First Edition 2022

ISBN 978-0-6482161-3-1 (eBook)
ISBN 978-0-6482161-5-5 (Paperback)
ISBN 978-0-6482161-6-2 (Hardcover)
ISBN 978-0-6482161-2-4 (Audiobook)

www.outcomeleaders.com

CONTENTS

INTRODUCTION

Jay's Story

Jay was vice president for an Enterprise Resource Planning (ERP) software vendor. He received a call from one of his largest customers, an automotive component manufacturer. It was not good news.

The customer told him: 'We're not achieving the operating metrics required by our parent company. We have decided that a new ERP system provides the best way to achieve those metrics. So, unfortunately, we're going to move away from your system.'

Jay knew the customer was not using the software well. The natural reaction would have been for him to convince the customer to relook at how they were using the software. And to show the customer how expensive and disruptive changing software could be.

But Jay did not fall into that trap. He knew that once a customer has decided to switch software, it's hard to change their mind.

Jay was aware the parent company was putting pressure on the local president to achieve the operating metrics. The parent company wanted a clear plan for how that would happen. In response, the president had chosen to change ERP systems.

Jay met with the local president. He did not mention software. Instead, Jay talked about operating effectiveness: 'We have consultants who understand what it takes to achieve operational effectiveness. They understand all aspects, not just the technology. Let me send in a team to review your operating practices. They'll write a detailed report with everything needed to achieve and exceed your company's operating metrics.'

Jay handed over an example report, one the world's largest consulting firms would be proud of. Jay's team submitted a proposal to run the study. And included a price tag of US$100,000.

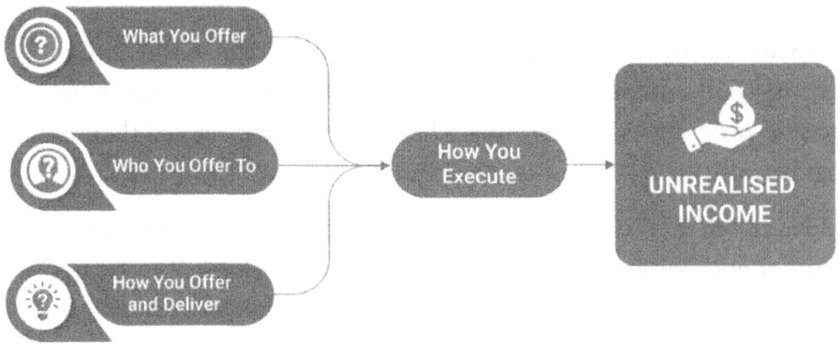

Figure 1: The structure of this book

Book 1 begins with Why Adopt an Outcome Program. It then covers the three topics on the left of the diagram. We start with What You Offer customers. The foundation of your offering is the core outcome you enable, called a Success Outcome. This is an ongoing business result which the customer's higher management regard as a success. It's achieved through four to eight smaller outcomes, called Contributing Outcomes.

This very simple structure underpins your whole outcome program.

Next, we consider Who You Offer To. We'll introduce you to the concept of a Decider. They are the person most likely to start business initiatives that need your offering.

We then discuss How You Offer and Deliver. We'll introduce you to the six decisions a customer must make so you (and they) succeed. As they progress through their customer journey, they make a decision at six key points. If they make the right decision, their journey with you continues. If not, it stops. We call these the IDEEAL Decisions.

Decision-making is first an emotional commitment, followed by a rationalisation. Each of the six decisions requires a different emotion. We discuss each of the six emotions required.

With an understanding of the six decisions and the emotion needed for each, you'll be ready for your customer engagement model. In designing your model, there's only one principle you need apply. How do you elicit each of the six IDEEAL decisions? The answer is an Outcome Lifecycle with six steps.

Six decisions. Six emotions. Six steps.

Book 2 covers the execution element. We consider how you'll use your outcome approach to create the six decisions.

We'll also talk about what this means for your offering.

And we will talk about how you can proceed with an Income from Outcomes program.

How to Read the Book

Book 1 provides the concepts that underpin the Income from Outcomes program. It's around 90 pages, with lots of diagrams. I'd recommend reading all of Book 1.

Book 2 deals with how to execute the concepts. When launching an outcome program, we recommend that you don't take a big-bang approach. An outcome program should be an evolution, not a revolution.

Give some thought as to how you want to get started. We offer some pathways in Part 14 (How to Proceed with an Outcome Program). Concentrate on the parts of Book 2 relevant to how you want to proceed. For example, if you want to use outcomes to improve lead generation in existing customers, you should consider:

- Part 1 – Outcome-based Lead Generation
- Part 6 – Outcome Techniques
- Part 7 – Segmenting Your Existing Customers
- Part 14 – How to Proceed with an Outcome Program

Eventually, your outcome program should be pervasive. All customer-facing departments, and product development, should adopt your outcome approach. But achieving this to-be state is an evolution, not a revolution.

In time, you can read the whole of Book 2. You'll find ideas from one area spark your thinking in other areas. If the book serves its purpose, in time it will be dog-eared and filled with notations or Post-it Notes. If it's an eBook, you'll have highlights and notes littered throughout.

As an author, I like nothing more than seeing a book I've written in this condition.

I hope *Income from Outcomes* helps you on your own outcome journey. And helps you tap into Unrealised Income.

BOOK 1 – THE CONCEPTS

Section 1 – Other Challenges Addressed by an Outcome Program

Higher Purpose

Companies feel growing pressure to stand for something. To have a purpose that transcends selling products or services. But it often seems supporting a bigger purpose will cost them financially.

And staff share the desire to serve a higher purpose. They want their work to tie to something meaningful.

There's a link between higher purpose and financial performance. In his Harvard Business School article 'Higher Purpose and Financial Performance', George Serafeim said: '... based on approximately 500,000 survey responses of worker perceptions about their employers ... We document that firms exhibiting both high purpose and clarity have systematically higher future accounting and stock market performance' (Serafeim, G., 2019).

Serving a core customer outcome offers clarity and creates a strong sense of purpose. And there's a much clearer link between the work your staff do and customers' benefits.

Present-Dayism

Customers that focus on solving current problems suffer from present-dayism. Business practices and markets change. And the speed of change increases all the time. A focus on current problems is too slow. It requires the customer to:

- Identify the problem
- Analyse it
- Decide how to solve it
- Investigate solutions
- Implement a solution
- Wait for the benefits to flow

All this takes time. And by the time customers have solved the original problem, new problems have arisen. Customers struggle in catch-up mode without ever getting ahead. That's frustrating for them. And limits their ability to produce breakthrough results.

Customers need to look ahead, not focus on current problems. Your outcome program will help them envisage a future to-be state. You'll then help them get there. They can get ahead. You need to help your customers overcome present-dayism.

Do-Nothing Decisions

For vendors, do-nothing decisions are painful. The sales team engages, applying all the resources needed to prosecute a deal. And then, the customer or prospect decides not to proceed. They don't buy from anyone.

It's expensive and wastes everyone's time. You need a more effective method for minimising do-nothing decisions.

New Buyers

For the technology industry, IT departments no longer dominate technology buying. In LinkedIn's seventh annual survey of technology buying (LinkedIn, 2020), respondents were asked: 'Which function has the most influence on your company's technology purchases?'

Sixty-three percent nominated departments other than IT. Line-of-business executives play a much stronger role.

This has two implications. First, your front-line teams must engage these line-of-business executives. But what do they talk about? Executives aren't interested in your products or services. They delegate evaluation of products and services.

Second, line-of-business executives will have a stronger focus on business results than traditional IT departments.

You need to arm your teams with topics and insights executives will value.

Servitisation

Companies in several industries are switching from selling their products to offering their products as a service.

Manufacturing Industry

The manufacturing industry has shown a strong trend towards servitisation. Many manufacturers no longer sell their products. Instead, customers pay for outcomes achieved. For example:

- Rolls Royce Aero no longer sells jet engines (Akhter, 2018). Instead, it charges for the number of miles flown. It takes responsibility for ensuring engines are flight ready. The more miles flown, the better it is for the airline. And that drives higher revenue for Rolls Royce.
- Michelin Fleet Solution moved to an outcome-based metric in its price per kilometre (PPK) model in 2001. The company now charges truck fleets based on mileage per month rather than the number of tyres purchased. Michelin introduced maintenance, delivery and free replacement services with the model. The company shifted its focus from manufacturing quality tyres to improving transportation outcomes (Balaji, S).
- Philips Lighting switched from selling LED lamps to lighting-as-a-service for Schiphol Airport, Amsterdam. It sells light and ensures the system functions properly. The airport uses energy-efficient lamps to reduce consumption (IEDP, 2017).
- Nicklin Transit Packaging shifted its focus to selling the safety of the packages as an outcome, with a remote monitoring service. It sends real-time information about the location and safety of packages to customers (Baines, 2020).
- Baxi moved away from selling boiler systems to selling heating as a service. Rather than pay for units of energy (kWh), customers buy hours of warmth in their home, called Warm Hours

(Catapult Energy Systems, 2020). Baxi ensures customers get the contracted number of Warm Hours, regardless of weather conditions.

- Hitachi Rail is building new trains in the United Kingdom. Hitachi owns and maintains the trains, and the Network Rail System pays Hitachi for 'on time services' (Gill, J. et al., 2021).

Research by analyst firm IDC suggests 82% of manufacturing companies are moving towards servitisation. In 2018, it was still early days, with fewer than 5% having fully transitioned. But more than 60% were already in the early stages (Nebuloni et al., 2018).

PA Consulting Group conducted a study in 2019. They observed that 85% of manufacturers aim to deliver output-based services in three to five years. And 57% aim to deliver outcome-based services. It also found companies realised 15% to 35% growth margin on services. This was much higher than that attained on their products (PA Consulting, 2019).

Pharmaceutical Industry

There's a move away from payment based on number of pills consumed to the efficacy of drugs for the population. For example:

- Germany's Act to Reorganise Pharmaceuticals Market in the Statutory Health Insurance System (AMNOG) uses the health-outcome approach. The framework develops contracts based on the clinical benefits of drugs (Lauterbach et al., 2016), while the Institute of Quality and Efficiency in Healthcare (IQWiG) researches the clinical benefits and outcomes of new drugs.
- In the US, the Patient-Centered Outcomes Research Institute (PCORI) also studies new drugs (Lauterbach et al., 2016). This is part of an expanding movement towards an evidence-based and patient-oriented approach in the US drug industry.

- Australia's Pharmaceutical Benefits Scheme (PBS) has adopted a similar approach. For example, they developed a contract with the manufacturers of the drug Bosentan based on outcomes (Owen et al., 2008).
- New Zealand's PHARMAC regulates drugs and vaccine contracts. Health outcome is one of the main considerations (Babar, 2015).

Technology Industry

In the technology industry, pricing based on results enabled is not new. Software company i2 provides one of the better-known examples. Founded in 1988, they offered pricing linked to the results their supply chain management software enabled.

We would note you don't need to adopt outcome-based pricing to use an outcome approach. But as pricing moves from subscription-based to consumption-based to results-based, having an outcome approach will position you well.

Customer Experience

Customers buy products or services to achieve a business outcome. If they don't achieve that outcome, the vendor has a serious problem. As does the customer.

A vendor might optimise every touch point with a customer. They might provide intuitive, friendly systems. And they might have conquered multi-channel engagement. Customers might love engaging the vendor. But if the outcome isn't achieved, someone with authority will demand change. And the vendor will lose the customer.

So, your primary concern should lie with enabling a business outcome which the customer's higher management will regard as a success. Everything else is secondary.

Section 2 – Unrealised Income

The other challenges addressed by an outcome program help. But tapping into Unrealised Income provides the strongest reason for adopting an outcome program.

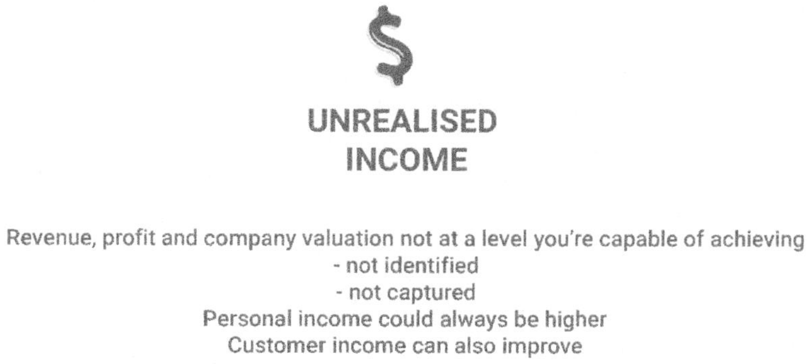

Figure 3: Unrealised Income

Few businesses would feel they could not create more revenue, earn more profit or improve the company's valuation. Perhaps churn is higher than you'd like. Or expansion revenue isn't where you expect. Or New Business (New Logo) revenue isn't what you'd hoped. Or margins aren't at the right level.

These issues point to Unrealised Income. Unrealised Income exists when revenue, profit and valuation are not yet at a level you are capable of achieving. Even businesses performing well feel they could do better.

A Side Benefit

Maximising financial performance for your company should benefit your career. That will increase your personal income. For most executives, increasing the share price also creates wealth. And who wouldn't like their personal income to be a bit higher?

Customers

You'll enable a core business outcome which your customer's higher management will regard as a success. And that will improve their revenue and profit as well.

So, Unrealised Income applies at the company, personal and customer level.

Four Types of Unrealised Income

The first is revenue you could go after but haven't yet identified. It might be new revenue sources from existing customers. Or it might include new markets.

The second lies with revenue you know about but aren't as effective at capturing as you'd like.

The third lies with reducing costs.

Finally, company valuation:

- Increasing revenue and profit will increase company valuation.
- For software companies, the move from on-premise to software as a service (SaaS) creates much higher valuation multiples. As discussed earlier, Boston Consulting Group and Geoffrey Moore point to the adoption of outcome programs. An outcome program opens the door to even more effective pricing models. And these may also drive higher valuation multiples.

You'll see as the book unfolds how an outcome program addresses Unrealised Income.

Un-Problems

Eight critical problems combine to create Unrealised Income. We call these the Un-Problems, for reasons which will become obvious.

Un-Problem # 1 – Unengaged

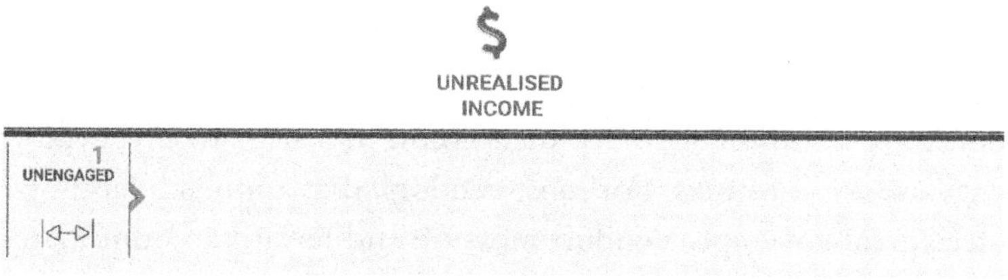

Prospects and customers no longer engage vendors early in their buying process

Figure 4: Unengaged

Customers and prospects have changed the way they buy. They now engage vendors much later in their buying cycle.[(1)] This has caused the influence a vendor has on the buying process to plummet.

Early in the buying cycle, a prospect or customer makes decisions that will impact the potential revenue for a vendor. They might, for example, decide not to move forward at all. Do-nothing decisions abound.

Or they might decide they don't need your products and services. Perhaps a different approach will fulfil their needs.

Given the importance of these decisions, vendors would love to engage early.

But what can you do to re-engage customers and prospects? Right now, they don't want to talk with you in the early stages.

Un-Problem # 2 – Unexpanded

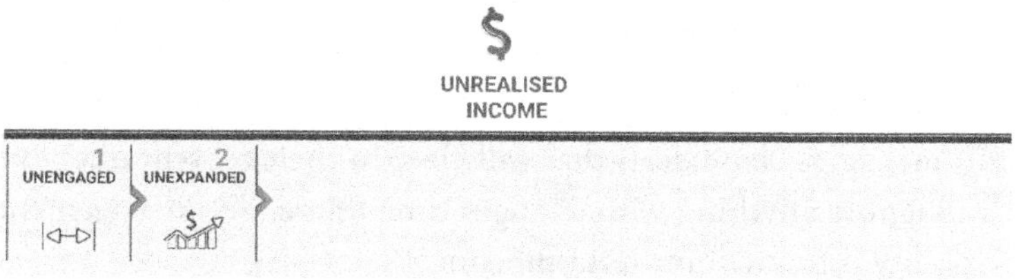

Expansion revenue program lacks a disciplined approach with a consistent, repeating focus

Figure 5: Unexpanded

You might feel expansion revenue from existing customers (upsell and cross-sell) isn't as high as it might be. This Un-Problem has a few facets:

Poor Coverage of Existing Customers

Vendors have traditionally relied on account managers to cover the existing base of customers. For most vendors, that approach has not produced great coverage. Vendors measure and reward account managers for new deals. If account managers have enough deals in their pipeline to make quota, they don't chase other customers.

A lack of coverage of these other customers means vendors miss:

- Customers planning to leave—by the time the customer notifies, it's too late for corrective action. As a result, churn is higher than it should be.
- Opportunities for a deal—possible deals slip through the cracks because no one knows an opportunity exists.

Lack of Executive Engagement

Sales managers exhort their account managers to reach their customers' executive level. But once there, what do they talk about? The customer's executives don't want to talk about products. They delegate product analysis to other people. And account managers often feel they don't have enough business knowledge to have a general business conversation with a customer executive.

Passive resistance ensues. Account managers promise their manager to establish executive contact, but it never seems to happen. Instead, they continue with their usual lower-level contacts.

Short-term Pipeline

Account managers chase deals that will close in their current quota period. Most will ignore anything with a longer time frame. So, management has little visibility into medium-term pipeline.

Lack of a Solid Foundation for Regular Engagement

A strong expansion program needs disciplined, systemic engagement. And a strong business driver for that engagement. It's not enough to have 'check-in' meetings. Or periodic business reviews that focus on usage of your products or services.

Stagnant Customers

Some customers drift along year after year, not changing. As a result, their spend with a vendor stagnates. You need a way to activate these stagnant customers.

These challenges result in expansion revenue being less than your company is capable of achieving.

Un-Problem #3 – Unscaled

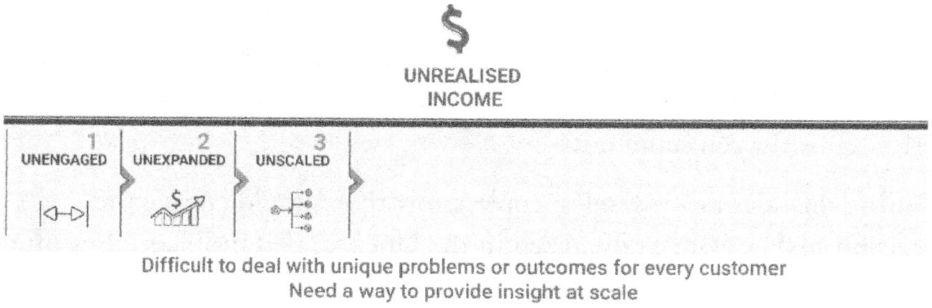

Figure 6: Unscaled

Customers and prospects think they are unique. They want solutions for what they think are unique problems or methods to achieve their unique outcomes.

They expect vendors to come up with an approach for them alone. That is both time-consuming and expensive for vendors.

You need to provide unique insight to prospects and customers—including business insight if you are to engage at the executive level. The challenge lies with what insight you provide. And being able to do that at scale.

Un-Problem #4 – Uncompetitive

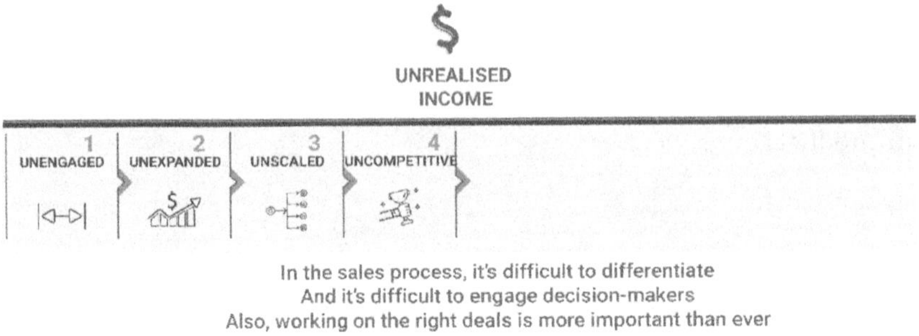

Figure 7: Uncompetitive

Your competitors have good products and services, possibly as good as yours. They also have good sales teams and know how to show their products can solve customer problems. They have an impressive list of customers, including references and case studies. They've probably got an ROI calculator.

In short, they are a lot like you—meaning prospects and customers find it difficult to tell the difference. Which often drives a race to the bottom on price. How do you stand out?

During the sales process, salespeople know they should engage the decision-makers. But as discussed in the Unexpanded problem, they often hesitate and don't get to the executives as they should.

That drags down your win-rate.

At the same time, subscription revenue becomes more important. It used to be that if you won a deal, you made a profit. You received most of the revenue upfront. Not with subscription pricing. Now customers pay over time. The customer acquisition cost (CAC) may exceed revenue in the first contract period. If the customer cancels their subscription, you lose money.

It has become more important to engage with customers with the highest chance of success because they will renew their subscription, which allows you to make a profit. You need an approach to increase win-rates in competitive deals, and to win the right deals.

Un-Problem #5 – Unaligned

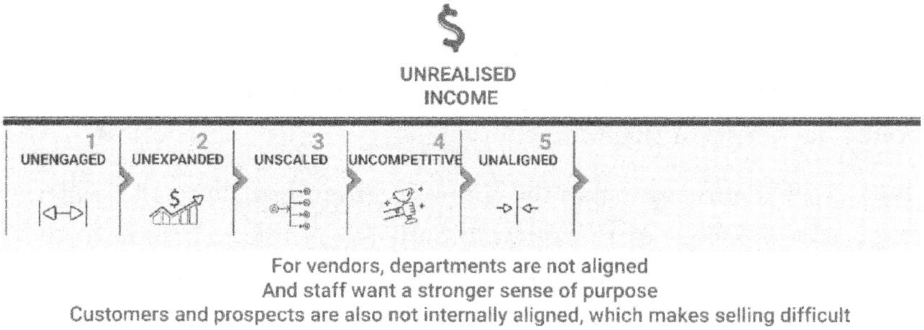

Figure 8: Unaligned

This Un-Problem has a few aspects.

Vendor Alignment

Each department within a vendor strives to do their best. They develop a plan to support the company strategy. And they work hard to make it happen. But often, alignment presents a challenge.

Marketing develops messages and campaigns to create awareness.

When sales engages customers, they don't always speak the same language as marketing. For sales, it's all about winning the deal.

Sales then hands over to onboarding or professional services. This group just wants to get the products live.

Then Customer Success gets involved with its focus on retention and product adoption.

Support cares about fixing the products if they break.

And product development faces a battle given each department has competing ideas for the product roadmap.

Channel Partner Alignment

You have a strategy and message to the market. But your channel partners aren't always aligned with you. This makes it hard for the market to have a clear perception of what you do.

Customer or Prospect Internal Alignment

A customer or prospect usually has many people involved in a buying decision. While the sales team tries to get alignment between these people, this often proves difficult. Each person has their own self-interest and view about what's best for their company.

This lack of alignment makes the common ground smaller and smaller. That causes the scope of a business initiative to shrink. Which in turn drives down the price the customer or prospect wants to pay.[2]

A lack of alignment affects:

- The internal functioning of a vendor
- The engagement with channel partners
- The employee experience
- The ability to sell deals

Un-Problem #6 – Unvalued

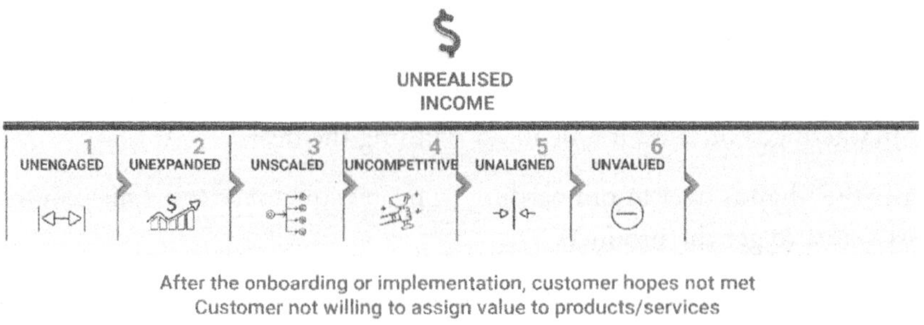

Figure 9: Unvalued

By the end of the sales process, the customer or prospect has a set of hopes for what they want to achieve. But after go-live, they don't always get what they hoped for. This can happen for a few reasons. First, the customer or prospect misunderstood what was possible. They had expectations no one could meet.

Second, the project objectives were vague. The customer did not know quite what to expect. But when they got to the end, they felt disappointed.

Third, they were oversold—the sales team positioned your product's capabilities a little more favourably than they should.

Fourth, the project itself ran into problems. When this occurs there's almost always contribution from both vendor and customer. But it doesn't matter who is at fault. If the customer doesn't feel successful, it's likely to cost you revenue.

Another challenge lies with the customer agreeing to the value of your products. You'd love your customers to say you helped them achieve a financial benefit. But they are often reticent to put a value on your contribution.

You need to maximise the chance of the customer feeling successful. And to be able to measure the results.

Un-Problem #7 – Unsung

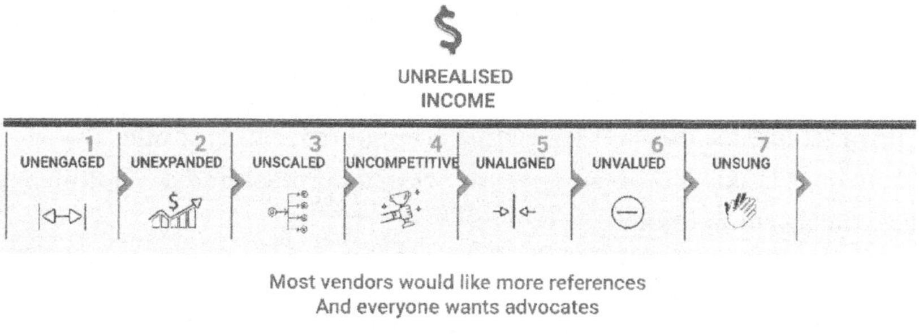

Figure 10: Unsung

Sales organisations can't operate without references. Sales needs friendly customers who, when asked, will say positive things about the vendor. But many vendors rely on only a handful of references they use again and again. An honest self-evaluation may reveal most customers as lukewarm.

Advocates proactively tell other people how great you are to work with. They're far more valuable than references. Unsolicited feedback to a potential buyer has strong credibility.

You need a program to increase references and advocacy.

Un-Problem #8 – Unmet

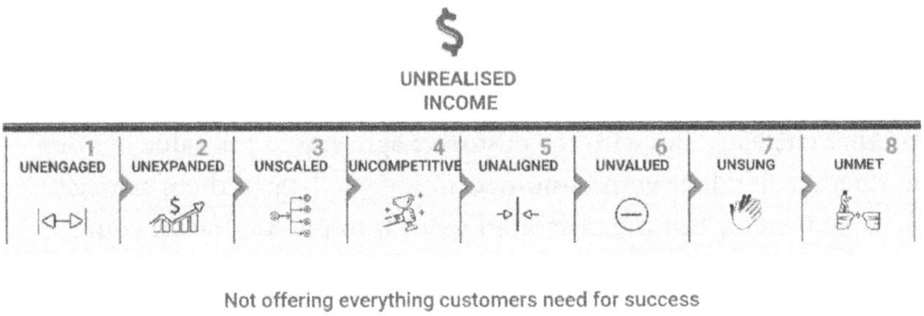

Figure 11: Unmet

Everyone has an opinion about how the product development budget should be spent. And those opinions rarely align.

Marketing relies on research and feedback to form views about what's needed in the product; the sales team wants to meet demands from larger customers; professional services wants to improve implementation; Customer Success gets feedback from a broad range of customers; support wants bugs fixed; and of course, product management and development have their own opinions.

These differing perspectives all have merit, making it hard to find a single lens with which to view them all. This can result in a less-than-ideal product roadmap. One that is often not focused on what will most help customers.

You need a way to align the product roadmap and agree on priorities.

Summary

Eight problems combine to create Unrealised Income. An outcome program addresses each of these problems.

And that will allow you to tap into your Unrealised Income.

Section 2 – What Are Outcomes?

We will start with some concepts introduced in *The Outcome Generation.* They provide a foundation on which we will develop the Income from Outcomes approach.

There's Always an Outcome

Whenever there is a commercial transaction of any type, there's an outcome. For example, if you buy a hamburger, you want to be full. That's an outcome. If you buy a cinema ticket, you want to be entertained. If you buy a sports car, you probably want to be noticed.

Figure 13: There's always an outcome

Two Types of Outcomes in B2B

We deal with two different types of outcomes in B2B environments. The first is a Product Outcome, which is the direct benefit of using your product or service. The second we call a Success Outcome, which is the bigger outcome the customer is trying to achieve. Your products are a means to this bigger outcome.

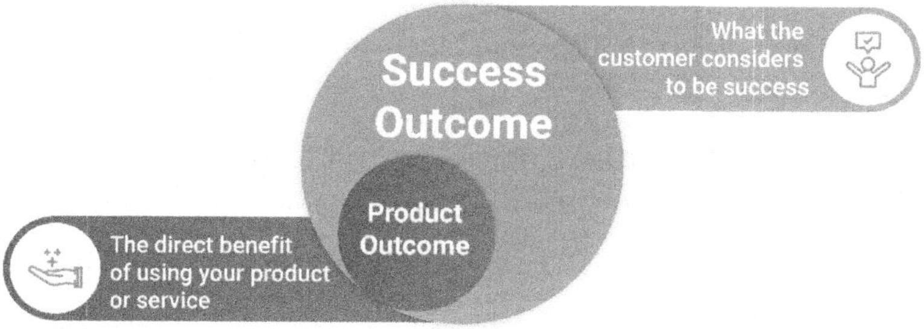

Figure 14: Product Outcome vs Success Outcome

The Product Outcome plays a part in the customer achieving the Success Outcome, but it's not enough by itself.

Imagine you go to a hardware store to buy a drill bit. You don't want to own a drill bit for its own sake, you want to drill a hole. If the drill bit works as it should, you will have a hole of a specific diameter. You paid for the drill bit, it worked as it should, and you have a hole. That's a Product Outcome.

But who just wants a hole? What we wanted to do was hang a picture. When we can step back and admire our handiwork, we've had a Success Outcome.

We did not buy a drill bit to get a hole in the wall; the drill bit was a means to an end. We wanted a framed picture hanging on the wall, looking great.(3)

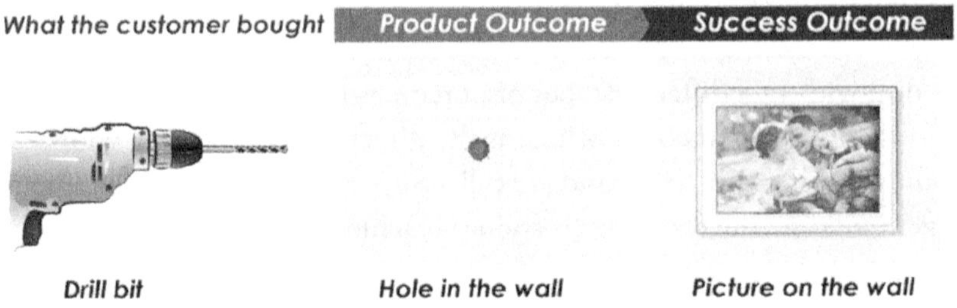

Figure 15: The hole in the wall vs the picture on the wall

Making a pie provides another example. Imagine you sell the ingredients for the finest pie crust.

But people don't buy pie crust ingredients just to make pie crust. They buy the ingredients so they can make a pie.

Making a great pie crust is a Product Outcome. But enjoying a delicious pie is the Success Outcome.

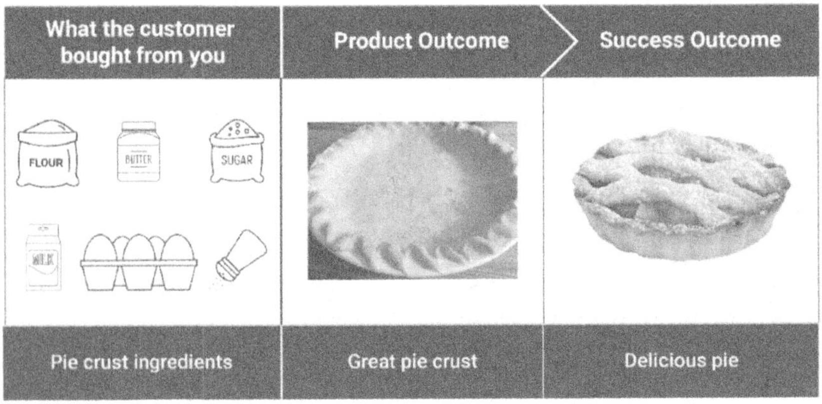

Figure 16: The pie crust vs the delicious pie

Section 3 – Why Distinguish Between Product Outcome and Success Outcome?

You have just hung a picture on your wall and step back to admire your handiwork. But it looks terrible. You've hung it in the wrong place.

The next time you have handiwork to do, you might hire a handyperson. And not go back to the hardware store for more equipment to do it yourself.

You won't spend more money at the hardware store. The drill bit you bought worked and delivered the Product Outcome of a hole in the wall. But you did not achieve the Success Outcome because the picture did not look great. The failure to achieve your Success Outcome will cost the hardware store future revenue—through no fault of its own.

Similarly, let's consider the pie. Imagine you have served your friends or family a slice. They have a look of expectation on their faces. They taste the pie, and their faces tell the story. The pie tastes awful. The pastry was great, but something else went wrong. It's a disaster!

You were expecting plaudits. Everyone tries to be polite, but you feel terrible.

The next time you're entertaining family or friends and you need a pie, you will probably buy one from the bakery. The seller of the pie crust ingredients will lose future revenue, even though the pie crust was perfect. The Product Outcome was achieved but the Success Outcome wasn't.

In B2B, these situations occur all the time. Your products or services work as they should. But the customer doesn't get the other elements required to achieve their Success Outcome. It's easy for the customer to blame the products or services rather than accept responsibility themselves.

That puts your future revenue at risk. Lack of success means the customer may cancel their subscription. Or reduce the number of seats or licences. Or just stagnate and not grow their investment with you. Each will cost you money.

It's the Success Outcome, not whether your products or services performed, that determines future revenue.

Figure 17: Future revenue is driven by the Success Outcome

It's in Your Interest to Enable the Success Outcome

There's a simple message from these examples. You should be crystal clear about the Success Outcome you help customers achieve. And do everything possible to help them achieve that Success Outcome.

To continue the pie example. You sell the ingredients for great pie crust:

Figure 18: Pie crust ingredients

But for someone to make a great pie, they not only need the ingredients for the pie crust. They need other ingredients. And they need tools. And there are tasks they must perform.

The entire range of what's needed to make a pie might look like this:

Figure 19: Everything needed for a delicious pie

The difference between the Product Outcome and the Success Outcome are all those other elements the customer needs to get right.

It's in your interest to ensure the customer gets them right.

Software developed on the Pendo platform automatically creates insight into product analytics and user behaviour. Ben Carey, Chief Customer Officer of Pendo, said:

Our customers write software to help their customers achieve a business outcome. If the end customer doesn't achieve that business outcome, the end customers will stop using the software our customers wrote.

Our customers need to know if the end customers are on track to achieve the business outcome. Our platform helps our customers identify and track key indicators of the end customers' business outcomes. And to take action to close any gaps.

There are two other ways we help our customers enable business outcomes for their customers. First, our customers can conduct retroactive analysis to find usage and behavioural indicators of the end customers' success. They can develop more refined indicators of likely success over time.

Second, they can create projections of future to-be states for their customers. Our customers can help their customers anticipate their business situation in, say, 90 days' time. All based on data collected automatically.

We see our role as much more than a product platform. We see our role as helping our customers create business outcomes for their customers. When that happens, everyone wins.

Section 4 – Characteristics of a Success Outcome

A New To-Be State

Whenever a customer achieves an outcome, they create a new to-be state or end-state. Your Success Outcome describes this to-be state. It's a new level of performance for the customer.

Ongoing

The new level of performance is not a one-off. The customer will want to maintain and even improve the new level of performance you helped them reach.

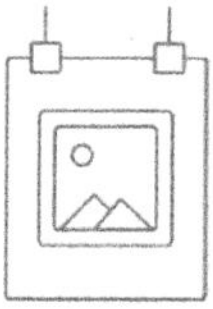

Travel Poster

Imagine a travel website's photos of great holidays—perhaps a pristine ski field; or a golden beach with a lucky couple sipping pina coladas; or an exotic city with people exploring the sights. You think, 'I want to be in that picture!' Your Success Outcome should create the same reaction for customers or prospects: 'I want to be in that picture!'[(4)]

A Phrase or Tagline

Describe the to-be state you help your customers achieve using a single phrase of 10 syllables or fewer. It will invoke the travel poster idea and create a mental image. It will act much like a tagline. And it will convey the core outcome you create for your customers.

No Jargon or Product

A Success Outcome should have no jargon, no

abbreviations and no industry terms. It does not include words or a phrase that represent a current fad. It also has no reference to product. A Success Outcome describes a to-be state, a needed business result. It does not say how you will enable the outcome.

Enduring

A Success Outcome describes an ongoing business result a customer's higher management would regard as a success. If you choose this business result well, in 20 years your customers will still want to achieve and improve that ongoing business result. The way you help customers will change, but the Success Outcome will be the same.

Promise

Your Success Outcome makes a promise to prospects and customers: 'If you work with us, this is what we will help you achieve.' It sets an expectation your competitors will find challenging to match.

Insight

The phrase you use provides insight by itself. It describes a to-be state. And many people will not have thought of what they need to achieve in those terms.

Examples of Success Outcomes

There's no right answer for the Success Outcome you serve. Two companies with a similar offering may choose different Success Outcomes.

These factors may influence your choice of Success Outcome:

- Your current skills and expertise
- The capability of your current products
- Which problems you believe are most significant to customers

- Your place in the competitive landscape
- Your strategic direction

Consider these possible Success Outcomes for different vendors:

Enterprise Resource Planning (ERP) Software

Companies buy ERP software to help them run their operations effectively. A Success Outcome for an ERP vendor could be 'Effective Operations'. This passes the test for a Success Outcome in the following ways:

- Ten syllables or fewer.
- To-be state—the phrase creates a mental image any senior executive could relate to and would want to achieve.
- Significant—an ERP vendor would be the most significant vendor for the outcome of Effective Operations. So, a customer needing to improve their operations would seek help from their ERP vendor.
- Enduring—ERP customers will always need effective operations. The way the ERP vendor enables this will change over time. But the Success Outcome won't change.
- No jargon—Effective Operations are business words any executive could understand.

Marketing Automation Vendor

Marketing departments understand they should create pipeline. So, they would subscribe to a marketing automation service to increase their pipeline. But a marketing automation vendor who understood the concept of Success Outcomes would have a different idea.

When CEOs approve marketing automation systems, they don't want more pipeline—they want more sales. If the pipeline increases but sales revenue doesn't, the CEO will consider the marketing automation system a failure. And may decide to cancel the subscription.

A clever marketing automation vendor would not focus on pipeline alone. They would focus on pipeline that sales could convert to deals. So, the Success Outcome would not be pipeline, it would be 'Closable Pipeline'.

They would work with customers' marketing departments to create Closable Pipeline. For example, they would help marketing and sales define the ideal prospect. They may then broker a service level agreement. Perhaps they will define what's required before marketing hands a lead to sales. And then what sales will do once they receive the lead.

For the customer of the marketing automation vendor, this results in an increase in sales revenue—delivered through Closable Pipeline.

IT Service Management Tools

Companies have a digital environment comprising servers, networks, applications, communications, mobile, security and so on.

IT Service Management vendors assist IT departments with digital environments. They offer infrastructure management, application deployment, support desks, security, analytics, IT operations, mobile and more.

For CIOs, the digital environment presents a no-win situation. Everyone expects their IT service to work well. If it does, the CIO gets no credit. If it breaks or is slow, the company will blame the CIO. It's a headache.

CIOs therefore want a smooth, reliable digital environment.

A vendor of IT Service Management Tools might choose a Success Outcome of 'Digital Environment Humming'. They offer a to-be state in which the digital environment runs smoothly and reliably. That would take away the headache for the CIO, who would pay a lot for that outcome.

Section 5 – Expressing a Strong Point of View

Two companies provide great examples:

Mercury Interactive

Founded in 1989, the company was in the business of software quality testing.

In 2000, cofounder Amnon Landan wanted to expand. He hired a new chief marketing officer (Christopher Lochhead). And began expanding its offering to monitor and optimise software at every level of an enterprise.

According to the authors of *Play Bigger* (which include Christopher Lochhead), the company needed to make its products make sense with each other and tell a larger story.

The company coined the phrase Business Technology Optimization (BTO). And created a point of view: 'Your business is running on software, and if the software is not performing well then the business is not performing well.'

In 2002, they had $400 million in revenue. By 2006, they surpassed $1 billion. Hewlett Packard bought them that year for $4.5 billion. The head of HP's Enterprise Business, Ann Livermore, said that BTO was half the reason they bought the company.

Source: *Play Bigger,* Ramadan, A. et al., 2016

Qualtrics

Qualtrics began in 2002 as a single-product market

research company serving academia.

The company went through several iterations leading up to a key decision in 2017. They committed to a point of view that companies needed a single platform for employee, customer, product and brand experience. And they coined the term Experience Management, creating a new category.

In January 2019, their success led SAP to acquire them for $8 billion: 'the largest private-enterprise software acquisition in history'. At its IPO two years later, the opening price valued Qualtrics at three times that amount.

Source: *Harvard Business Review*, 'The Founder of Qualtrics on Reinventing an Already Successful Business'. Smith, R. July 21.

Key Message

While the experience of both companies offers many lessons, both did one thing. They expressed a strong point of view about what they do for customers. And they adopted a short phrase to capture the essence of that point of view.

They said, quite clearly: 'Here is the one thing we do. If you need that, we are the only company to speak with.'

And their financial success followed.

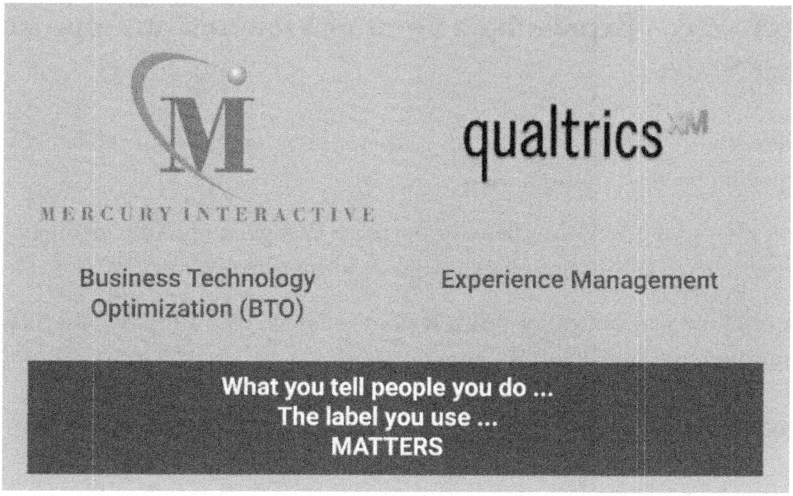

Figure 20: The label you give yourself matters

Section 6 – Expressing a Point of View and an Approach that Scales

Mercury Interactive and Qualtrics provide great examples of the benefit of expressing a clear point of view.

You also need a way to scale your outcome engagement with customers. We discussed earlier how asking your customers doesn't work.

Here's how you can express a clear point of view. And simplify and focus your outcome-based customer engagement.

Only One Success Outcome

For each line of business in your company, you need only one Success Outcome. If you only have one line of business, your company only needs one Success Outcome.

What do we mean by a line of business? A line of business is that part of your company which serves customers with a common set of problems.

Your Success Outcome captures the essence of what you do for your customers—it's your guiding light, your North Star. It will drive interaction with your customers. Eventually, all customer-facing departments will play a role in enabling the Success Outcome. It will also guide your product development.

You may find the idea of defining your business through a single core outcome unsettling. Doesn't that limit the range of customers you can engage? What if prospects or customers don't like the outcome you choose? Won't that damage your revenue opportunities?

Here is some context for this idea of a single Success Outcome.

Outcome Hierarchy

A traditional organisation chart of your larger customers might look something like this:

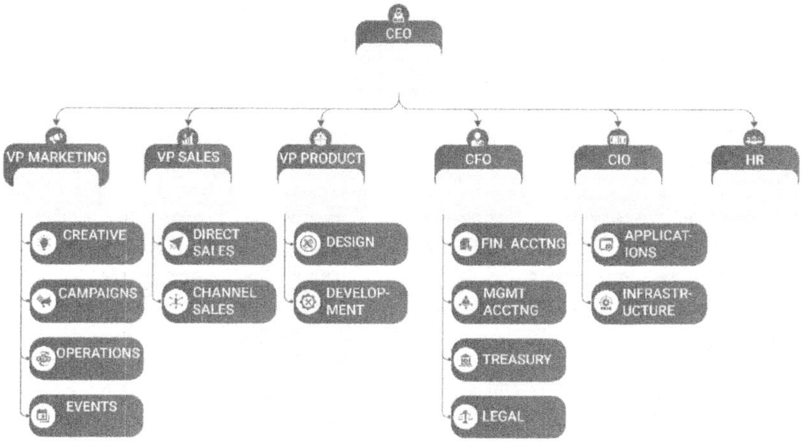

Figure 21: A customer's organisation chart

Organisations typically define each role by responsibilities.

But we can also think about each role in terms of outcomes—a single core outcome each role must deliver for the organisation. For example, the CEO delivers stakeholder returns; the Vice President of Marketing delivers pipeline; the Vice President of Sales delivers sales revenue; and the Vice President of Product delivers ready product.

At the top level, the organisation chart may now look something like this:

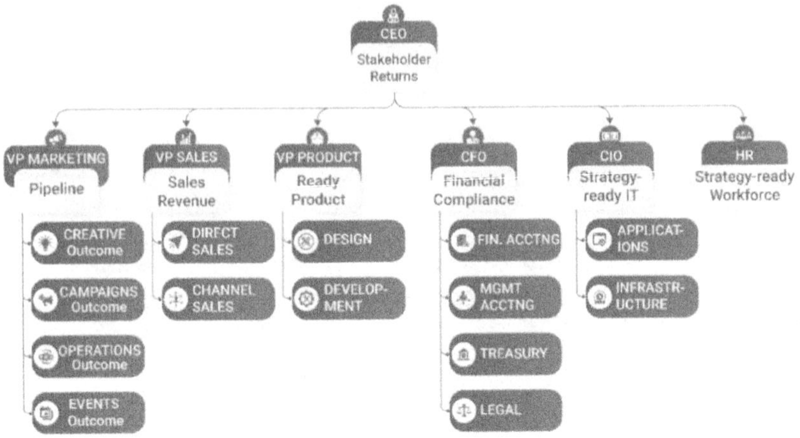

Figure 22: Outcome Hierarchy

An Outcome at One Level Is Enabled Through a Series of Smaller Outcomes at the Level Below

If everyone at the VP level delivers their outcome, it's likely the CEO will deliver stakeholder returns. An outcome at one level in the organisation is enabled through a series of smaller outcomes at the level immediately below.

Outcomes cascade from the top down. For example, consider marketing:

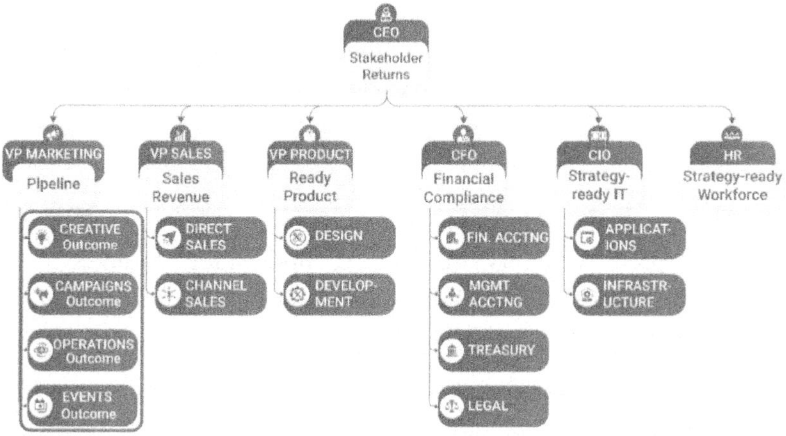

Figure 23: The outcomes at one level are achieved by a series of smaller outcomes at the level below

If the people immediately below the VP of Marketing deliver their outcome, it's likely the needed pipeline will result. The VP of Marketing will deliver their outcome.

Finding Your Place in the Outcome Hierarchy

We have introduced two ideas about outcomes. First, for each line of business as a vendor, you only serve one Success Outcome. Second, in a customer's organisation, each role delivers a single outcome.

Let's marry those two ideas. You should identify the role with the strongest alignment to your Success Outcome. It may sit near the top of the organisation chart, or halfway down, on the left, or on the right.

Significance

You need realism and a little humility in choosing your place in the Outcome Hierarchy. You may, for example, claim to increase your customers' profit. That may be true. But if that increase is only a small part of a customer's profit, the customer CEO (who owns profitability) won't come to you for help.

You should align with a person who will come to you for help, for whom you are the most significant external provider. When it's time for the customer to improve the outcome, you are their natural choice.

Decider

In B2B, customers undertake business initiatives. They don't get approval to buy your products or services. They get approval for a business initiative with a business case. Your products or services form part of the business initiative.

For you, the most important person within a customer is the person who starts and drives a business initiative. A business initiative that needs your products and services.

We call this role the Decider.

In the Outcome Hierarchy, you need to identify your Decider. This will be the person you align with and focus on.

We'll cover the role of Decider in more detail in Part 3, Section 2 of Book 1.

What If Lots of Roles Use Your Products?

You may provide a product or service used across your customers' businesses. For example, everyone uses email. If you provide email, the person owning the outcome most relevant to you won't be on the operational side of your customers' businesses. Someone in IT will own the outcome of ensuring the email service works. If you're the email provider, that's where you align.

Summary

The Success Outcome you choose for a line of business provides your guiding light, your North Star. In a single phrase of 10 syllables or fewer, you capture the essence of what you enable for customers. You become known in the market for your ability to enable that one outcome. And you align everything in a line of business around enabling that one outcome.

Section 7 – What If Other Companies Copy Your Success Outcome?

For a line of business, you've selected a single Success Outcome. It should be unique. You alone will offer this outcome to your market.

But what if competitors copy your Success Outcome? The news is good. It's likely you will have created a new category. And you're the first one to market in this new category. You are setting the pace.

Now imagine that you become the category king. You become the Facebook (now Meta), Google, IKEA, Apple or Amazon of a space you have defined.

In *Play Bigger* (Ramadan, A. et al., 2016), the authors say: 'Category kings, the data show, usually eat up 70% to 80% of the category's profits and market value.'

So, if competitors copy you, welcome it. And double down on your Success Outcome to become the category king.

Section 8 – Your Outcome Structure

You've now identified your Success Outcome. You have a point of view on the one thing your line of business does for customers. And you know where your Success Outcome fits in your customers' businesses.

Now we apply the principle that an outcome at one level is enabled through a series of smaller outcomes at the level below. You can identify these smaller outcomes. We call them Contributing Outcomes. Together, they contribute to achievement of the Success Outcome.

Typically, you'll find four to eight Contributing Outcomes for a Success Outcome. Think about a typical span of control in an organisation chart.

You now have your Outcome Structure. If you have, say, five Contributing Outcomes, it will look like this:

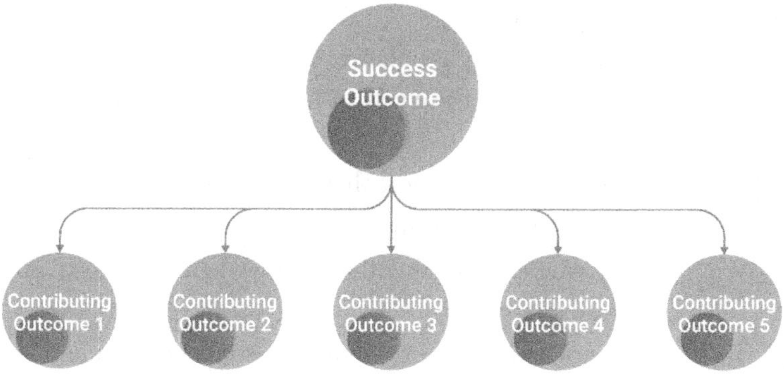

Figure 24: The Outcome Structure

For a line of business, this simple diagram captures the essence of everything you do for your customers. If your business has only one line of business, it captures your whole business. Everyone in the business can learn this simple diagram.

Marketing messages and campaigns will link to this Outcome Structure. Sales campaigns will use this structure to differentiate and to win business. Onboarding and professional services will focus on enabling this Outcome

Structure. Customer Success will use this structure to drive a repeating cycle of engagement with customers. Customer support will play a role, particularly in SaaS businesses. And this simple diagram will become the reference point for your product roadmap.

An Example of an Outcome Structure

Imagine a travel and expense management software vendor—a fictional company which we will call SCT Travel and Expense Management, Inc (SCT).

Travel and Expense Management Challenges

SCT knows most of their customers face the same challenges with travel and expense management.

Travellers, for example, might want:

- Choice—they want to:
    - choose loyalty programs that suit them
    - fly on airlines and stay in hotels that give them loyalty points
    - use ground transport that suits them
- Convenience—they want to:
    - fly when it suits them, not only when the lowest airfare is available
    - check into and out of hotels at convenient times
- Ease—they want:
    - easy, simple travel bookings
    - ease of submitting travel expenses
    - expenses paid on time

Customer management wants:

- Travellers to comply with the company's policies
    - Costs to be as low as possible within the policy—if the policy allows someone to fly business class, the company still wants the lowest practical business class fare

The needs of travellers and management often conflict, and that creates the major travel and expense management problems.

Success Outcome

SCT's leadership ran an exercise to consider these problems, asking: 'What if our customers' top travel and expense problems disappeared? How would we describe the to-be state the customer then enjoyed?'

They came up with: 'Customers would have satisfied travellers who believe a fair balance had been struck between their personal needs and the need to minimise travel costs.

'Management would be happy with a program that kept travel costs to a manageable level. They would also have insight into the level of compliance before individuals travelled. And that would allow for corrective action, or approval as an exception. Finally, the program would not result in an increase in staff turnover or dissatisfaction.'

The exercise helped align SCT leadership's thinking about the outcome they should create for their customers. They then debated the Success Outcome—a phrase of 10 syllables or fewer to capture the essence of the to-be state.

They chose: *Satisfied, Compliant Travellers.*

And this became SCT's Success Outcome. Everything SCT did now focused on creating a to-be state for customers of Satisfied, Compliant Travellers.

Contributing Outcomes

SCT then considered the Contributing Outcomes, the smaller outcomes needed to create the to-be state. For SCT's customers to have Satisfied, Compliant Travellers, their customers must:

- Have a clear, well-understood travel and expense policy, with both travellers and management accepting the policy.
- Provide travellers with as much choice as possible, compliant with the policy.

- Identify non-compliance and allow for intervention or approval as an exception.
- Enable timely submission of travel expenses with least possible effort for the traveller.
- Ensure the timely approval and payment of expenses.

They then captured each outcome as a phrase of 10 syllables or fewer:

- Acceptable Policy in Place
- Valid Travel Choices Maximised
- Exceptions Managed
- Easy Expense Submission
- Expenses Paid on Time

Achieving these five smaller outcomes would create Satisfied, Compliant Travellers.

Outcome Structure

SCT's Outcome Structure now looks like this:

Figure 25: The Outcome Structure for SCT Travel and Expense Management

This simple diagram became SCT's North Star—all their engagements focus on helping customers achieve Satisfied, Compliant Travellers by enabling the five Contributing Outcomes.

Section 9 – What It Takes to Achieve an Outcome – PSPSPS

Achieving any business outcome requires five elements:

- Processes
- Systems
- People Skills
- Partners
- Supporting elements

We summarise these as PSPSPS.

There's nothing new in this structure. Companies have understood the need to focus on these core elements for a long time.

The Five Elements

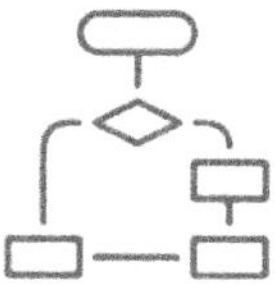

Processes

Companies need well-designed processes that lead to achievement of an outcome. Processes tend to evolve over time. Clarity around an outcome provides a great lens through which to examine current processes. And make changes if needed.

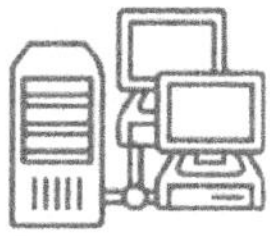

Systems

We tend to associate systems with technology. But systems can also be manual. For example, manufacturing companies used to keep inventory information on handwritten cards. That was a system. Even technology can create a 'manual system'. People use spreadsheets, for example, when the company's main system proves cumbersome. Or when they can't see a good reason to change away from spreadsheets.

Systems should create a single version of the truth and give the right people access when they need it.

People Skills

This heading suggests people trained in how to perform their role. But there's more to people skills in achievement of business outcomes. Clarity about why you ask people to perform a role in a certain way helps them, particularly if something out of the ordinary occurs. People should also understand how their work affects others. No one works in a bubble.

When making business changes, an effective change management process increases the chance of success.

And having a sense of higher purpose increases motivation. And that increases discretionary effort and reduces staff turnover.

Partners

Most companies have external partners, such as suppliers and external contractors, that play a role in enabling outcomes. Companies need to integrate their partners.

Supporting elements

You need policies that guide the way the company works. You need measurement and reporting focused on outcomes. You may need different compensation systems to reinforce the outcomes. These all represent examples of supporting elements required to achieve business outcomes.

Achieving a Business Outcome is Like Making a Pie

Earlier, we discussed the difference between a Product Outcome and a Success Outcome. We used the example of a pie. You might sell the ingredients for great pie crust:

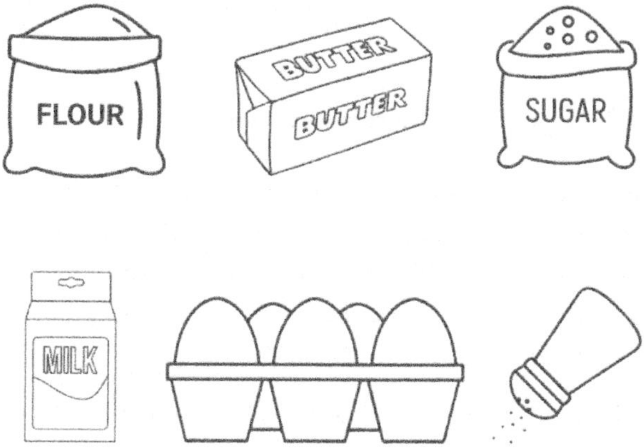

Figure 26: Pie crust ingredients

But to make a delicious pie, the customer must get all these elements right:

Figure 27: Everything needed for a delicious pie

PSPSPS is the business equivalent of all the elements needed to make a great pie.

Defining PSPSPS for Your Success Outcome

To achieve a business outcome, your customers need to get PSPSPS right. If the customer gets any of the elements wrong, the business initiative may fail. And as we discussed earlier, that can cost you future revenue, even if it's not your fault. B2B companies have an incentive to ensure their customers get PSPSPS right.

You should define PSPSPS for the Success Outcome you enable. You should have clarity about the things your customers need to get right. And you should aim to help them with as much of PSPSPS as possible. It's in your self-interest to ensure the customers get all PSPSPS elements right.

You can avoid making this complex. Identify the:

- Major processes
- Key systems
- Important people skills
- Main partners
- Significant supporting elements your customers need

The full list for a Success Outcome will probably consist of 12 to 18 PSPSPS elements. It will look something like this:

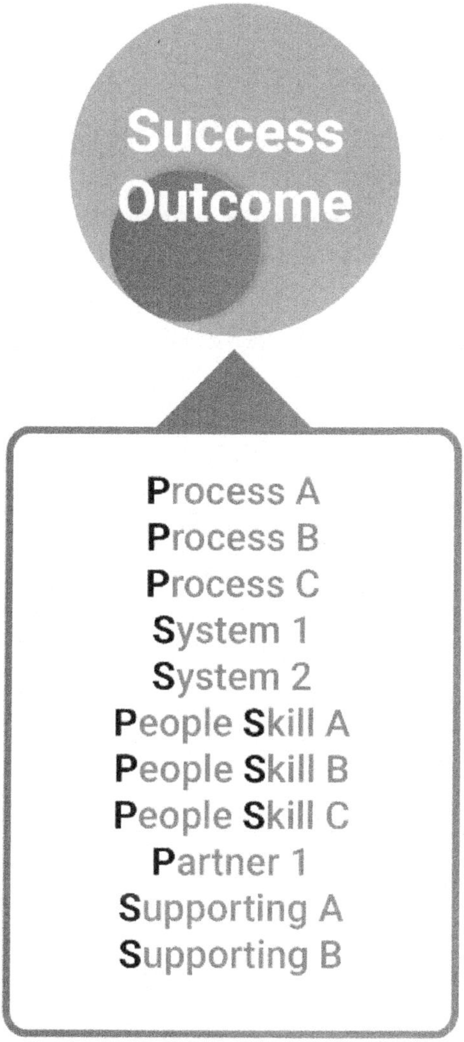

Figure 28: PSPSPS for your Success Outcome

Section 10 – The Benefits of Adopting This Outcome Structure

Your Outcome Structure has two elements:

- The Success Outcome you enable—the core outcome you create for your customers
- The Contributing Outcomes—the smaller outcomes that together contribute to achievement of the Success Outcome

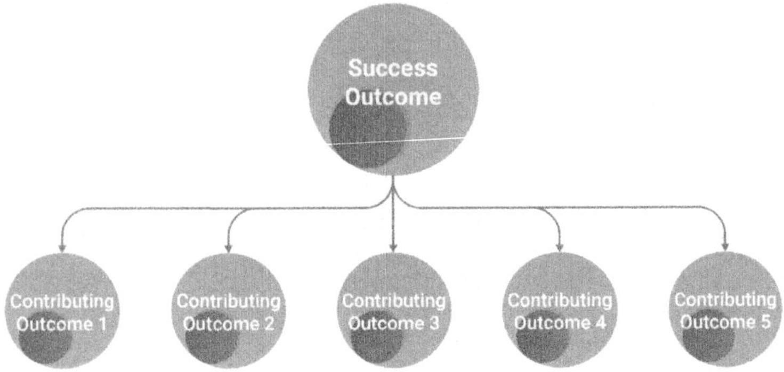

Figure 29: Your Outcome Structure has a Success Outcome and Contributing Outcomes

We can group the benefits of using this simple Outcome Structure into three sets:

1. Financial benefits
2. Market positioning benefits
3. Internal benefits

Financial Benefits

Retention

If you help a customer achieve a to-be state higher management regards as a success (the Success Outcome), they will want to keep you.

Expansion

You will develop a repeating cycle of engagement with your customers. You will jointly focus on continuous improvement of the Success Outcome. With each engagement, you will identify a business initiative to improve the Success Outcome. These business initiatives will drive your expansion revenue.

New Business Win-rate

You will improve your win-rate for three reasons. First, you'll stand out from competitors—your Success Outcome will create a mental picture higher management in a prospect will want to achieve. And you will be the experts at enabling this outcome.

Second, you'll engage decision-makers (Deciders). Your ability to provide insight on a successful business outcome will get you to the executive level—and keep you there.

Third, you'll create more references and advocates. When those key customer people change companies, they will bring you into their new company.

Channel Performance

Your channel will have a clearer message about what you do and why customers and prospects should buy your products. It will also open new services opportunities for your channel, which will make it easier to recruit partners.

Margin

Experts can charge more. Think about the medical industry. General practitioners earn good money, but specialists make more. If you need heart surgery, you want a specialist—and you understand you will need to pay a lot for the heart surgeon's service.

The same will apply to you. You will become known as an expert in enabling the Success Outcome. And like the heart surgeon, you can charge more.

Medium-term Pipeline

Your focus on continuous improvement of the Success Outcome will drive your expansion program. It will identify short-term opportunities. But it will also identify business initiatives the customer isn't yet ready to start. They are opportunities for next year. Across your customer base, this will create a medium-term pipeline.

New Product Revenue Growth

When thinking about new products, your starting point becomes the Success Outcome: What's the best way of improving the Success Outcome you enable? This helps in two ways:

First, you're thinking about everything the customer needs to achieve the Success Outcome (PSPSPS). The next improvement may not be in your product—it might be a guide to best processes; or a change management tool; or a policy template. Your development will be driven by how you can best enable the next improvement in the Success Outcome.

Second, you will have a different view of disruptive technology. Normally, you might see a disruptive technology as a threat to your current products. But if it provides a better way to enable the Success Outcome, you might embrace it. Even at the expense of your current product line. Embracing innovation increases the likelihood of you prospering over time.

Market Positioning Benefits

Branding

You will become known as the best at enabling your Success Outcome. What you do for your customers will be crystal clear to everyone. No ambiguity. Any company that needs your Success Outcome will come to you.

Unique

Your Success Outcome defines a market you can focus on. And dominate.

Trusted by Customers

All engagement with your customers focuses on enabling or improving the Success Outcome. Over time, your customers trust your focus. As a result, you achieve a reputation of being a company that customers can trust.

Advocates and Community

You'll give credit for successful outcomes to your Decider and their team. They'll become heroes and heroines internally who will in turn become your references and advocates. Communities will form around them and the outcomes you enable for customers.

Internal Benefits

Simplicity

Your Outcome Structure forms the foundation of everything you do for your customers. A diagram with perhaps six circles captures everything you do as a line of business. All staff can understand and relate to this simple approach.

Alignment

Every internal department has the same clear focus. Each has a role to play in enabling the Success Outcome for customers. That creates alignment across your company around a single purpose.

Scale

You don't deal individually with customer problems or outcomes. You link the problems to your Outcome Structure. And you can then have the same conversation with each customer. This approach scales.

Sense of Purpose

Your Outcome Structure describes a clear purpose for a line of business. Everyone knows the one core outcome, the Success Outcome, you enable for customers. And this lifts everyone's sense of purpose. Selling and implementing a good product is a good thing. But creating a measurable business outcome for customers represents a higher calling.

Having this clear sense of purpose and higher calling will increase motivation. And that will improve recruitment, retention and discretionary effort.

Longevity and Continuity

Your Success Outcome describes a business need that will not change. For example, consider our fictional company SCT Travel and Expense Management. They help customers achieve Satisfied, Compliant Travellers. In 20 years, their customers will still want Satisfied, Compliant Travellers. The way SCT helps their customers achieve this to-be state will change. But the underlying business need won't change.

You will become known in the market for your Success Outcome. And in 20 years you'll still serve that same Success Outcome.

True Customer-Centricity

In every commercial transaction, there's an outcome in play. You'll focus on one core customer outcome—the Success Outcome. By aligning around this outcome, you will be truly customer-centric.

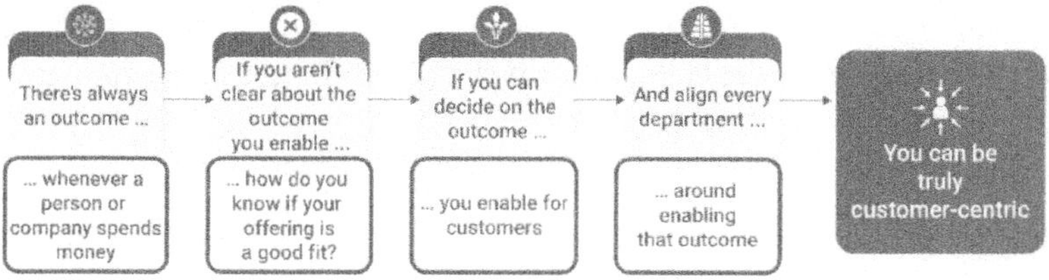

Figure 30: Truly customer-centric

Section 11 – Multi-step Outcomes

The Outcome Structure works well when engaging directly with business users of your products and services.

What happens when there is an intermediary? Perhaps you sell and/or implement through partners. Or your products assist an internal support department such as HR or IT to support their internal customers. In this case, the business model changes from the diagram on the left to the one on the right:

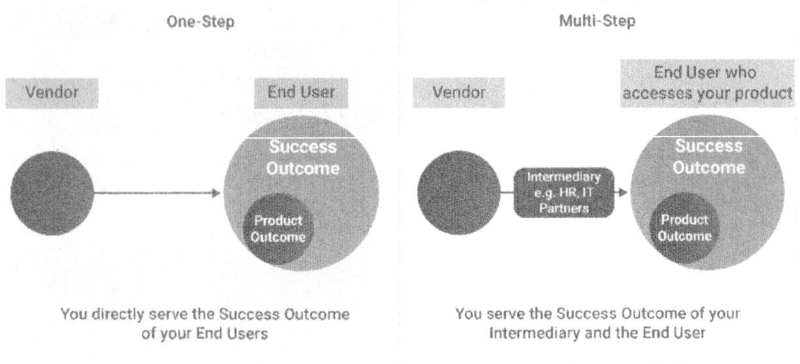

Figure 31: Multi-step Outcomes

Let's look at three examples of intermediaries:

Channel Partners

If you have a sales channel and/or services partners, you may not deal with the end customer using your products. You will need your channel partners to adopt an outcome approach. If you show them enough benefits, they will embrace this approach.

The core message lies with how they will benefit financially—and that is they will benefit in the same way you do. Help them understand the benefits and you'll get them on board.

Internal Support Departments Such as Human Resources (HR)

Internal support departments have many of the same characteristics as your partners. They act as intermediaries with the business end-users.

Consider a HR (Human Capital Management) system as an example.

The business case requires overall benefit for the company, not just how it benefits the HR department. Yet it's likely the HR department will start this business initiative. So, you need to convince the HR department of the benefit to them of adopting a company-wide HR system. And then you need to help them position the overall benefit to their company.

It requires both steps. HR must first buy into the idea of a new system and become enthusiastic supporters. But the HR department will often lack the ability to sell the benefit to the rest of the company. That's now your job to help them with the internal sales.

Broad-Based Tools Supported by IT

These include office suites provided by companies such as Microsoft or Google, email and meeting tools.

As with the HR system, business end-users will use these products. But Information Technology (IT) will lead any evaluation and provide first-line support. IT has a strong interest in meeting the needs of the business end-users. But they also have strong concern over the support they will get from the vendor.

So, vendors need both to create a great outcome for the business end-users, and then do the same for IT.

Summary

We have covered:

- The difference between a Product Outcome and a Success Outcome.
- For a line of business, your Success Outcome underpins everything you do with customers.
- You need to find where your Success Outcome fits in your customers' Outcome Hierarchy. And identify your Decider, the person most likely to start a business initiative needing your products and services.
- Contributing Outcomes enable the Success Outcome.
- Customers need the right PSPSPS to achieve their outcomes.

Extensis helps creative teams control creative chaos. Toby Martin, CEO, said: We have adopted an outcome structure to drive our business. It begins with our Success Outcome—Creative Chaos Controlled.

We help more than 140,000 customers across the world. And we can see how they can struggle to effectively manage digital assets such as pictures, videos and creative work product. This struggle creates chaos in a creative team. It drags down productivity.

We have a single aim as a company—to help creative teams achieve the to-be state of Creative Chaos Controlled.

We've identified five smaller outcomes needed to achieve this to-be state. Creative teams need:

- Creative Assets Stored
- Creatives Assets Organized
- Creative Assets Found
- Creative Assets Shared
- Creative Assets Controlled

We have an Outcome Structure that looks like this:

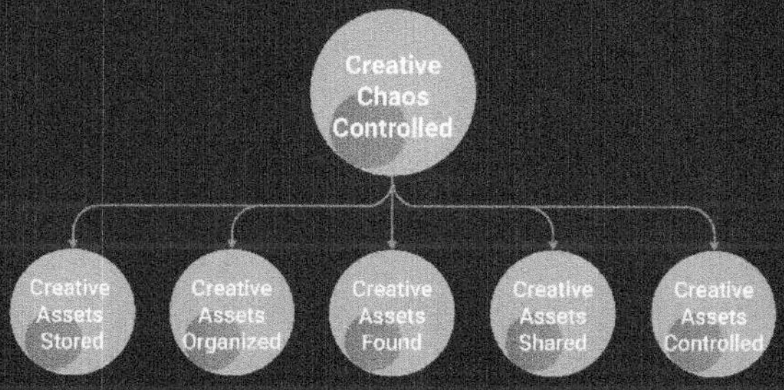

Figure 32: The Extensis Outcome Structure

This simple structure drives everything we do. It has created alignment across all departments. And it has created a common lexicon we all use.

PART 3 – Who You Offer To

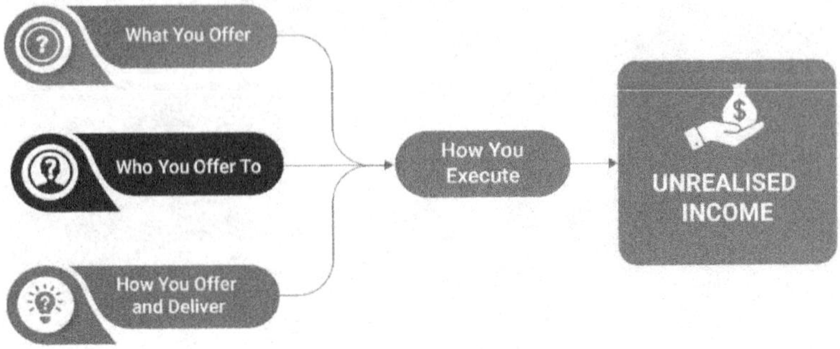

Figure 33: Who you offer to

What We'll Cover in Part 3

We'll begin with what influences people when they make decisions.

Then, we will consider the people you will engage in customers and prospects. This includes four key roles:

- Decider
- Approver
- Recommender
- Executor

We'll talk about the role each plays, with a strong emphasis on the key role—the Decider.

Section 1 – What People Consider With Business Initiatives

We've discussed how B2B companies acquire the bulk of products and services through a business initiative. When contemplating a business initiative, people will typically consider three things:

- What's In It For Me? (WIIFM)
- What's In It For the Company? (WIIFC)
- Risk

WIIFM – What's In It For Me?

Few people engage in a business initiative unless they expect a personal win of some sort. They'll consider 'What's in it for me to take part, or promote, or approve this business initiative?' Perhaps they will eliminate a problem that has bugged them. Or they will achieve recognition and possibly a promotion. Or they will earn a higher bonus.

A buying decision is like any other decision. It is first an emotional process followed by rationalisation. WIIFM will often drive that emotional element.

To increase revenue, any initiative you discuss needs to include a personal win. Ideally, you would be subtle in how you convey that personal win. You would not imply someone might act primarily for a personal win.

Once people have made an emotional decision, they then 'justify' the emotional decision with more rational reasons. That's where WIIFC comes in.

WIIFC – What's in It for the Company?

In B2B, any sizable investment requires formal approval, for which the Decider needs a strong business case. The return on investment will need to meet or exceed the minimum investment hurdles.

To increase your revenue, you need to ensure any customer initiative has a strong business case. WIIFC needs to be clear and strong.

Risk

Taking on a business initiative has risk for anyone involved. If the initiative fails, they at the very least get a black mark against their name. In the extreme, a failed initiative may cost someone their job and may also impact company results.

You need to help people achieve confidence in a business initiative. They'll want a plan which they feel has a high chance of success, as well as a great business case.

Section 2 – Decider

We've discussed the Outcome Hierarchy and the idea that each role in a customer delivers one core outcome. We suggested that you should align with the role whose outcome most closely fits your Success Outcome. This role is your Decider, the person who will focus on business results you most strongly influence.

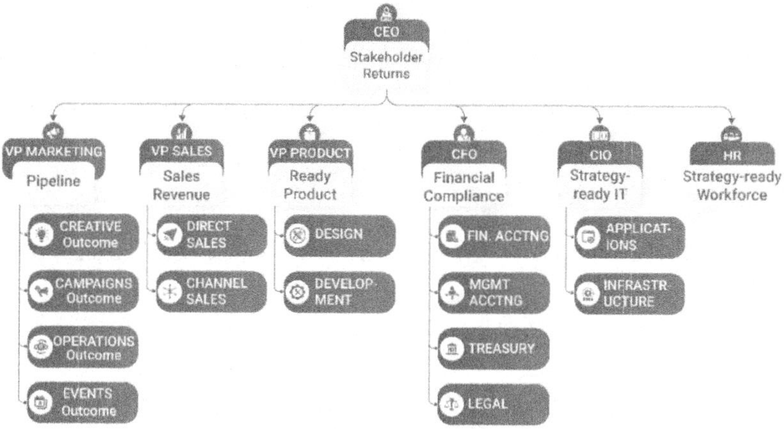

Figure 34: Outcome Hierarchy

Deciders are measured on the core outcome they own. They will start a buying cycle when they need to improve their results in this area.

For the Decider, your Success Outcome will have a significant influence on the core outcome they own. You and the Decider have a mutual interest … achieving and then improving the Success Outcome.

Six Decisions

In Part 4 of Book 1, we will introduce you to the six decisions you need customers and prospects to make. When the Decider Decides to Investigate ways of improving their results, they make the first of these decisions.

It's likely the Decider will also make the other five decisions you need the customer to make.

Identifying your Decider will prove key to the success of your outcome program.

Characteristics of the Decider

The major characteristics set out below provide a guide, not a rigid set of rules. They should help you identify the Decider in your prospects and customers.

Personally Invested

Your Success Outcome has a major influence on the core outcome they own. You'll make a big difference to whether they succeed.

Authority and Influence

The Decider will have enough authority or influence to drive each phase of the lifecycle.

Budget

They may not own the budget. They may need an OK at each phase from the person whose budget will fund the initiative. But they have enough influence to get that OK.

Personally Engaged

They'll spend time on the business initiative at every phase. For example, they won't delegate the development of a plan in the What to Do phase. They may delegate some activities, but they will want to drive the final plan. They will probably see the business initiative as the most important project they are working on.

Rule of Thumb

The Decider is:

- The most senior person in a customer ...
- ... who will devote significant personal time and energy to a business initiative ...
- ...linked to your Success Outcome

80/20 Rule

You will identify your Decider, the role most likely to start a business initiative that needs your products and services.

But sometimes other people will start actions that can lead to a business initiative. Perhaps the CEO or other senior executive decides that results need to improve in a specific area.

It's hard to identify every possible person and circumstance that might lead to a business initiative needing your products and services. And it would make your outcome program very complex.

Concentrate on the role that, in most cases, will start the right business initiative. If you're right 80% of the time, that's enough. You can build your program around engaging that role. You can deal with other circumstances as exceptions when they arise.

Educating the Decider

You've spent time analysing your customers and you've identified the Decider. They have a single outcome they must deliver for their company. But they may not have thought of their role in those terms.

In this case, you can give them insight to clarify and simplify their focus. You can help them see the importance of the one core outcome they own. And show how it aligns with your Success Outcome. You then have the foundation for collaboration with your Decider.

Going Above the Decider

Sales teams learn they should get to as high a level as possible within a prospect or customer. They should try to reach the economic decision-maker. They are told that if they engage someone who doesn't have budget authority, they are talking to the wrong person.

But this approach can backfire. And that's because it's more difficult to get the person above your Decider to act.

You've chosen where your Success Outcome fits in your customers' Outcome Hierarchy. You've identified your Decider, who needs your Success Outcome to succeed. Your Decider has a core outcome they create for their company.

Your Decider's core outcome is achieved through four to eight smaller outcomes at the level below.

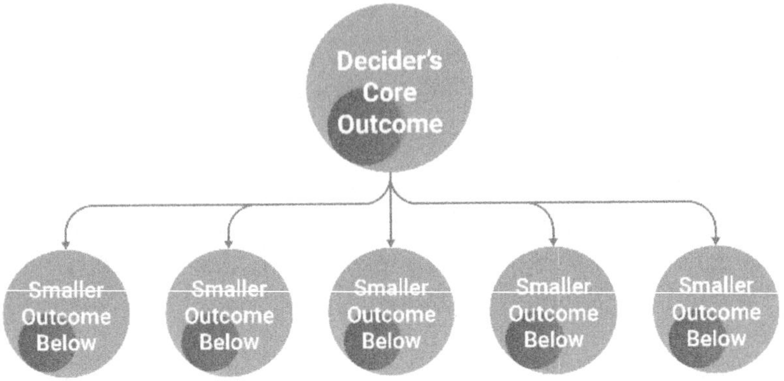

Figure 35: Your Decider's core business outcome

Your Decider's boss has a core outcome they enable for the company. It is also made up of four to eight smaller outcomes at the level below. Your Decider's core outcome is one of the boss's smaller outcomes.

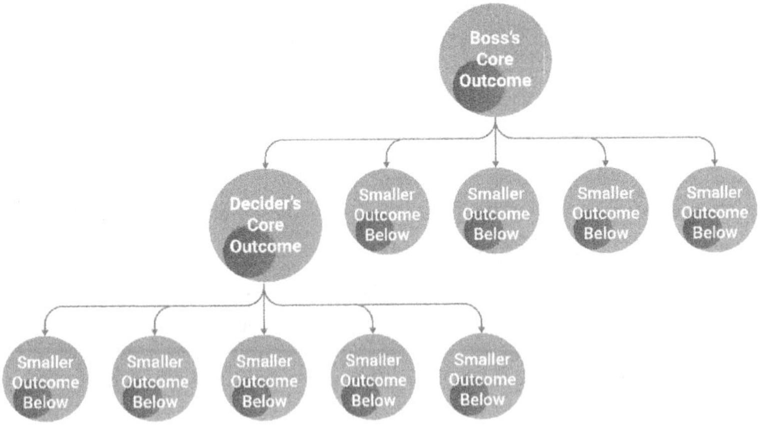

Figure 36: Your Decider's boss's core business outcome

Catalysing your Decider's Boss to Act

A buying cycle will only start when someone concludes their results need to improve. Your Decider's boss has four to eight outcomes competing for their attention. The boss might agree your Decider's outcome should improve. But not strongly enough to act. They have other outcomes to weigh up.

It's hard for your sales team to convince the Decider's boss. Your Decider has a much greater chance of success. Assume your Decider is a trusted lieutenant—they will have much more influence on their boss than your sales team.

Focus on Your Decider

So, instead of trying to reach the economic decision-maker, focus on your Decider. Get them to Decide to Investigate. It's much more likely you'll influence them to act.

Then arm them with the tools to influence their boss. Your Decider may need the boss's OK to move forward. But the Decider has a far greater chance of getting that OK than your sales team.

Which is not to say you should not engage above the Decider—but weigh up the purpose of meetings above your Decider. If possible, plan the meetings with your Decider.

What If Your Decider Won't Act?

It's possible your Decider won't act. So you might decide to go above the Decider to get some action. You contact the Decider's boss to get a meeting.

Think about what will happen after your meeting. The boss will ask the Decider's opinion. What do you think the Decider will say?

You might convince the boss that the Decider lacks competence or drive. But how often does that succeed? If you can't motivate your Decider to act, you have little chance of any business initiative going ahead.

Many Deciders

If your products and services impact many departments, you may need to engage more than one Decider. For example, each department head may have a strong say in whether an initiative goes forward.

But there will likely be a Lead Decider, someone who initiated the decision to investigate ways to improve results. They went first. Even if all the Deciders are peers, one will take the lead. For a given business initiative, that's the person you need to identify.

Examples of Deciders

Enterprise Resource Planning (ERP) Software

Companies buy ERP software to help them run their operations as effectively as possible. A Success Outcome for an ERP vendor could be 'Effective Operations'.

In larger organisations, the person most concerned with effective operations will be the Chief Operating Officer (COO). They are the most likely ERP Decider.

The CEO will care about effective operations. But the CEO is not the Decider because they would delegate the work of improving operations to the COO.

Marketing Automation Vendor

A marketing automation vendor should help customers create 'Closable Pipeline'. The person in a customer who owns Closable Pipeline will vary depending on the size of the customer.

In larger companies there'll be someone in marketing who owns lead generation. That person will be the Decider. In larger companies, the Decider will not be the chief marketing officer (CMO), who has a broader remit.

In smaller companies, the Decider will almost certainly be the CMO.

Travel and Expense Management

In our earlier fictional example, SCT Travel and Expense Management decided on a Success Outcome of 'Satisfied, Compliant Travellers'.

Larger customers will have a Travel Department, the head of which is the most likely Decider. In smaller companies, the Chief Financial Officer is the most likely Decider.

Bottom-up or Product-led Sales

Some vendors thrive with a bottom-up or product-led selling approach. Companies such as Atlassian, Slack, Dropbox, SurveyMonkey and Calendly provide excellent examples.

In this approach, an individual in a company decides to use a new product, perhaps for a project. That individual introduces other people in the project to the product. They'll use the free version making it easy for everyone to get onboard.

At some point, lots of people in that company use the product and it reaches critical mass. Someone in the company now raises the possibility of the product becoming a corporate standard. Or an enterprise salesperson approaches the Decider. The Decider Decides to Investigate the possibility of making the product a corporate standard.

For the vendor, a customer decision to standardise will create much larger revenue opportunities. And there is now a Decider involved.

In cases like this, the initial sales don't involve a single Decider, or even a Lead Decider. It's a groundswell effect. But the decision to adopt the product as a corporate standard will involve a Decider.

Services Companies

For product companies, identifying the Decider should not prove challenging. Services companies, however, can face a tougher challenge. For example, a consulting company may have an offering that any manager could use. Perhaps they offer change management consulting. Any manager driving a significant change would benefit from support by a change management consultant.

A change management consultant helps customers achieve a new to-be state. They could identify a Success Outcome they enable—possibly 'Change Made Successful'.

Any manager leading a business initiative may act as a Decider for a change management consultant. All of those managers need 'Change Made Successful'.

Summary

B2B companies need to know their Decider—they are the person most likely to decide results need to improve, in the area of the business you help. Until that decision is made, there's no buying cycle.

Section 3 – Approver

The approval process in a prospect or customer will end with an approving body. This body has the authority for final sign-off on a business initiative. Depending on the size of the investment, the approving body may be the board, an executive committee or a single executive.

In larger organisations, other people will need to sign-off before it reaches the approving body. These Approvers don't have the authority to approve the initiative—but they do have the ability to veto or delay it. Approvers will usually sit above the Decider in the Outcome Hierarchy.

But Approvers may also exist outside the direct 'chain of command' above your Decider. For example, initiatives involving technology may need sign-off by the IT department.

Getting Approval

When designing your outcome program, you'll focus on activating your Decider. But you'll also need to consider how to activate the Approvers.

Approving bodies will require a formal submission.

For in-between Approvers, you can help your Decider to craft the messaging required to activate these Approvers.

Approvers will also consider the three aspects we discussed earlier:

- WIIFM—What's In It For Me
- WIIFC—What's In It For the Company
- Risk

The approving body will consider several questions:

- First, is the business case for this initiative strong enough to approve?
- Second, how does this initiative compare to other initiatives? Most companies have more potential initiatives than they can fund. So, how does this initiative compare to others?

- Third, do other factors such as risk, timing, market considerations or regulatory influences affect this initiative?

The approving body will need a clear picture of the impact of this proposed initiative. They'll want to understand:

- The to-be state envisaged
- How the business will change
- How the initiative will run—how long it will take, how much resource it will absorb and who will be involved
- And of course, they will need a detailed business case

Customers aren't always great at putting together submissions for approval. They may not undertake these kinds of initiatives very often.

It's in your interest to ensure a submission has everything the approving body needs to make a positive decision. If the approving body sends it back with questions, you run the risk the initiative will lose momentum.

In Part 1, Section 9 of Book 2, we will discuss how to help customers and prospects develop high-quality submissions for a business initiative.

Section 4 – Recommender

Recommenders provide advice and guidance to Deciders. They help them feel confident to move forward with an initiative. They don't have the authority to say yes. But they often have enough influence to kill an initiative.

Under an outcome program, you don't sell to these people—you form joint working parties with them. Together you analyse potential initiatives and make recommendations to Deciders. Recommenders will often play a critical role in enabling the new business results. So, you want to form a collaborative relationship from the start.

Under an outcome program, what you focus on changes. With a traditional customer engagement model, you would focus on the problems the customer has. And then show how your products and services provide a solution.

Now, you will help the customer envisage a new to-be state. Achieving that to-be state will need more than your products. Any business outcome requires PSPSPS elements. You can help with some of those. But the customer will need to work on them as well.

For the business initiative to succeed, the customer must get all five PSPSPS elements right. This will only work through collaboration of the customer team and your team. Recommenders will form the backbone of the customer team.

Technical Recommenders

As well as business recommenders, the customer may have technical recommenders. IT departments can play this role. They will ensure the initiative complies with IT policies and that IT has the right resources to support the initiative.

It's common for customers to see any initiative involving technology as an IT project. Under an outcome program, this will change.

The old assumption held that getting a technology product live would produce the needed business results. You will help the customer understand why that is not true. A business outcome requires PSPSPS.

IT may play an important role in any initiative involving technology. But it's in your interest to help the customer see it's not an IT project—it is a business initiative. That increases the chance the joint project team will address all aspects required to deliver the business outcome.

Section 5 – Executor

Once an initiative has approval, you and the customer will form a joint project team. On the customer side, the people assigned will execute the initiative with you.

Some of the executors may have been recommenders. Others may be involved for the first time.

The executors will carry the bulk of the workload during the business initiative. You'll work with them throughout the initiative to deliver the business objectives set for this initiative.

Section 6 – The Place of Users

Some organisations will involve key users before approving an initiative. Obtaining user buy-in makes sense. If users rebel and won't adopt new business practices, the initiative will fail.

Together with the recommenders, you'll develop a plan for the new to-be state. And you'll plan what will change in the business.

It makes sense to socialise the new arrangements with users. They may identify challenges you missed. They may have suggestions for how you could do even better. Both would produce a better business outcome.

And the consultation, provided it's genuine, will increase commitment of the users to the new working arrangements.

Section 7 – Change Management

You will focus on the outcome throughout any business initiative. But the customer's staff must still adopt your products and the other changes in PSPSPS. If the staff resists, there's a good chance the initiative won't deliver the business results needed.

Change management plays a key role. Staff need to understand:

- Why the company wants to make this change
- How it will affect them personally
- What's in it for them (why they should support it)
- The to-be state envisaged
- What will need to change to get to this new to-be state
- How long it will take
- How you will measure success

When the initiative succeeds, you need to celebrate the success. You need to reinforce the continued commitment to the new approach. In summary, you need a change management program as part of the initiative.

Section 8 – Personae

It helps to have a clear picture of the people you engage in your prospects and customers. Creating personae provides an easy way to conceptualise those people.

We provide some ideas on how to set up your personae in Appendix 1.

Summary

We have touched on four key roles in a customer:

- Decider—decides business results should improve and has the authority to start an investigation
- Approver—signs off on expenditure for the business initiative
- Recommender—provides advice and guidance to Deciders
- Executor—delivers the business initiative

PART 4 – How You Offer and Deliver

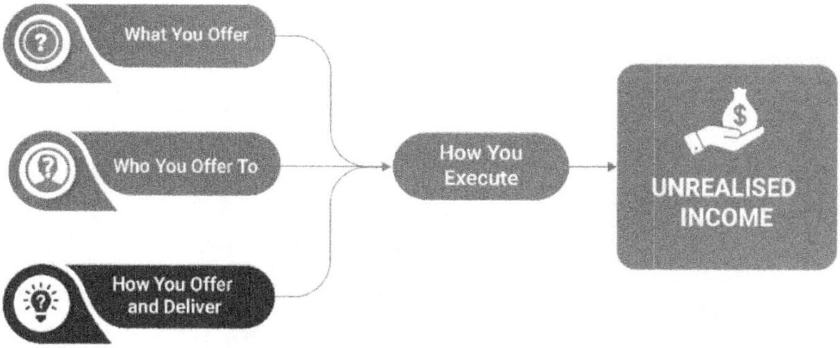

Figure 37: How you offer and deliver

The Rule of Six

Part 4 covers three important concepts, each with six parts:

Concept 1 – Six Customer Decisions

Prospects and customers undertake a journey. At each step, they make a decision before they proceed to the next step. If at any step they fail to make the required decision, the journey ends—at least for you.

You need clarity on the decisions you want customers to make at each step in their journey. It turns out there are only six. And they are the same for every customer.

Concept 2 – Six Emotions

Decision-making is first an emotional commitment to a path, followed by rationalisation.

If you want customers to make the six decisions, they need to experience the right emotions. It turns out there are six emotions needed.

Concept 3 – Six Steps

You know the decisions and emotions. You can now design your customer engagement to create the six emotions needed to elicit the six decisions.

You will create a step in your customer engagement model for each of the six decisions. Each of the six steps will be outcome-driven.

In summary, you need:

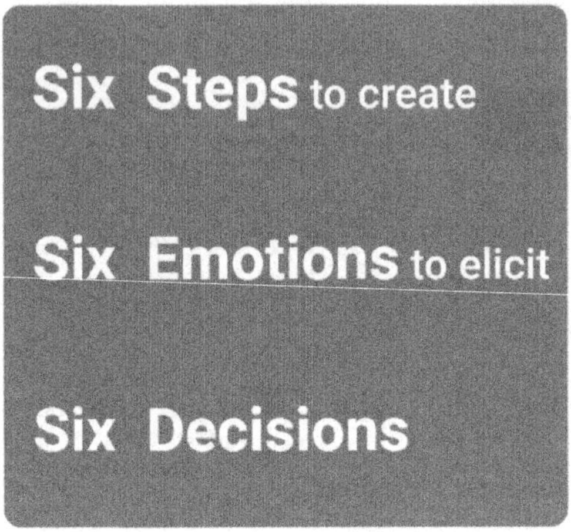

Figure 38: The Rule of Six

Section 1 – The Six IDEEAL Decisions You Need Customers to Make

You want the customer to make six decisions:

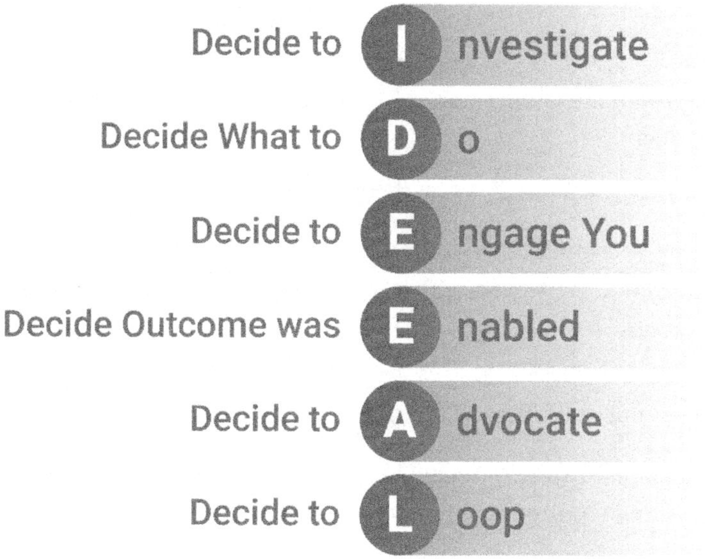

Figure 39: The six IDEEAL Decisions

The First Decision – Decide to Investigate

No buying cycle ever begins until someone with enough authority (the Decider) decides their results need to improve. And they Decide to Investigate ways to achieve that.

First, something must catalyse the Decider to consider their results. Perhaps it's the annual budget process; or benchmarking data you provide; or feedback from their boss that makes them feel their results need to improve.

Once they Decide to Investigate, the potential for a deal exists.

The Second Decision – Decide What to Do

It's a Business Initiative

Companies buy B2B products or services as part of a business initiative. They rarely submit a proposal to a board or executive committee for approval just to buy a product.

To gain a sale, you should ensure there's a clear business initiative that:

- Has a clear scope
- Describes a to-be state that higher management would consider a success
- Has a clear business case
- Sets out roles, responsibilities, timeframe and effort required

It's in your interest to ensure there's a well-documented business initiative with everything required for approval.

Type of Business Initiative

The Decider will now develop a plan for how to achieve better business results.

Better results don't always require technology. Any business outcome requires PSPSPS elements. It's possible any of these five elements could deliver better results. For example, improving processes might prove enough; or more staff training; or better integration of partners; or improved measurement and reporting.

Shape of the Business Initiative

In the past, customers and prospects engaged vendors early in their Decide What to Do phase. Vendors could provide insight on global best practices, what other companies were doing and what technology was available.

Now, companies find that information online. Vendors' ability to shape the business initiative in their favour has plummeted.

We will discuss how to re-establish that ability when we introduce the BuILD process in Part 1 of Book 2.

Salesforce helps companies achieve meaningful connection with their customers. Michael Kirschner, Senior Director, Customer Success Programs said:

We have an internal structure for planning called V2MOM. It stands for Vision, Values, Methods, Obstacles, Metrics. Leaders at all levels in Salesforce use this structure to plan how they will drive improvements in Salesforce's business. And it helps align everyone's plans.

V2MOM now forms part of a methodology we use to help our customers plan improvements in their business. We've learned the importance of ensuring the customer has a clear plan to achieve the business improvements they need.

Our methodology begins with the Listen phase. We seek to understand the customer's vision in the areas of the business where Salesforce can help.

We then move to the Build Trust phase. We use the V2MOM structure to create alignment between ourselves and the customer on the strategies they will employ and the outcomes we will jointly create.

In the next phase, Partner, we help the customer develop a plan to achieve the business improvements. The plan covers all aspects needed to achieve the business objectives. And just as V2MOM helps Salesforce align internally, our methodology creates a plan that aligns all the customer's affected departments, at all levels.

Once the customer commits to the project, we move into the Succeed phase. This has three steps: Implement, Realise Value and Grow. We concentrate on jointly enabling the business outcomes, then using that success as the foundation for future growth.

And the methodology works. In customers where we've used this approach, we're engaging at more senior levels. And we're seeing stronger alignment in the customer, particularly between business and IT.

The program has contributed to record low attrition in Salesforce. And that's critical, given 74% of our revenue comes from existing customers.

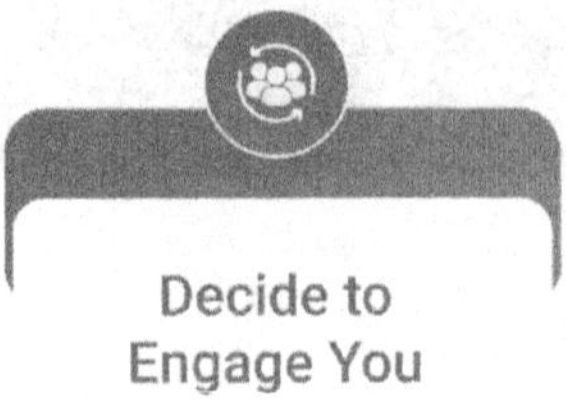

The Third Decision – Decide to Engage You

The customer or prospect now has an idea of how to achieve better results. They have a plan for the changes they will make in the five PSPSPS elements.

Now they will decide who they will engage to help them. They have a choice. And you want them to choose you. You have two aims during this phase.

First, you want to ensure their selection criteria favour you. You want to compete on your ability to enable the Success Outcome. You don't want this phase to be a product feature/function shootout. You want them to adopt outcome-focused selection criteria.

Second, you want their submission for approval to contain everything required. You don't want to be selected and then find the submission flounders. You want the submission to breeze through the approval process.

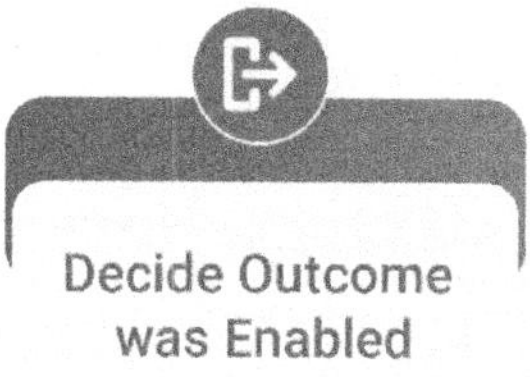

The Fourth Decision – Decide the Outcome was Enabled

We have already discussed the Un-Problem of Unvalued. Projects reach their end and the customer's hopes haven't been met. This can happen for all sorts of reasons:

- Expectations might have been poorly set during the sales process.
- The project scope was vague—the customer wasn't clear about what they wanted to achieve and is not excited by the result.
- The project had problems to which, usually, both customer and vendor contributed.

Vendors need the customer to declare success and report that the project enabled the outcome needed.

This requires a few things:

Clear Expectations

Before the project begins it must have clear, measurable objectives. Everyone involved must understand the outcome needed and what it will take to enable that outcome.

Outcome-focused Project Management

The project team must review progress towards the outcome in every Project Review meeting and Steering Committee meeting. If the project doesn't look like it will enable the outcome, both teams need to take corrective action.

Measurement of Results

You first need to establish a baseline 'stake in the ground' from which to measure improvements. This baseline measurement might be taken before the project begins, or before it goes live. It provides a starting point from which to compare post-project results.

At the end of the project, after allowing time for benefits to occur, measure the results. If you've delivered the objectives set at the start, everyone can declare success.

And the customer can Decide the Outcome was Enabled.

High Radius enables measurable business improvements in the Finance function, assisted by their AI and Autonomous Software. Pankaj Jagtap, SVP of Professional Services, said:

"We believe customer projects only succeed when they deliver a measurable business result for the customer - a result agreed before the project begins. And we live this belief.

That means, before we start a project:

- we set up the MASC - Mutually Agreed Success Criteria
- we ensure the MASC align with the Feasible Business Case mutually developed for this project
- we verify there are no client commitments from the sales stage not included in the Mutually Agreed Success Criteria

At key checkpoints we verify the project is on track to deliver or exceed the MASC. If not, we agree on corrective action.

Go-live does not end the project. The project continues until we verify we have achieved or exceeded the MASC.

And we reward the highradius team based on achievement of the MASC. That's what drives their bonuses.

We've learned that creating measurable business improvements for customers creates measurable business improvements for us.

The Fifth Decision – Decide to Advocate

Becoming an advocate isn't a trivial thing. If your customer declares success, the key players will probably be willing to act as references. This is a passive role—when you ask them to talk with another customer or prospect, they will do so.

An advocate proactively tells other people how great you are, without waiting for you to ask.

There is a simple way to create advocates and reinforce your references. First, you and the customer deliver measurable business improvements. Then, you ensure your Decider and other key project members get credit. By highlighting the great performance of this team, they will become heroes and heroines.

And heroes and heroines will become your advocates.

This is not something you can request—they will decide on their own. But by making them into heroes and heroines, you will have laid the foundation for them to Decide to Advocate.

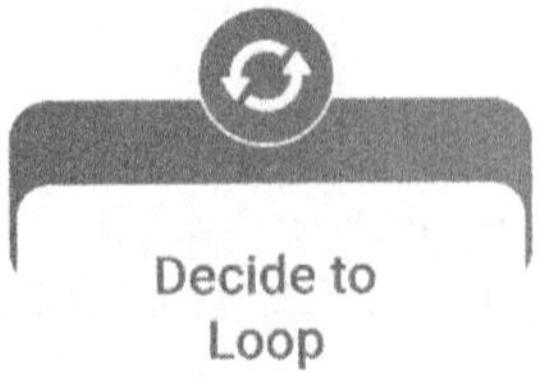

The Sixth Decision – Decide to Loop

At the end of the implementation, customers expect success—they expect the business changes to work smoothly and the benefits to flow.

Usually, that's as far ahead as they are thinking. They're not planning the next round of improvements.

A Different Perspective

Your outcome program will change that. Instead of seeing a business outcome as a one-off event, you'll encourage the customer to think of continuous improvement of the Success Outcome. A Success Outcome will never be 100% achieved. It's the nature of a Success Outcome. Opportunities for improvement will always exist.

Your outcome program will help the customer see the need for continuous improvement of the Success Outcome. And agree to a repeating cycle of engagement to deliver this continuous improvement.

The repeating cycle provides the foundation of your expansion program (upsell and cross-sell). You want them to Decide to Loop through this cycle time and again.

You will create a flywheel effect. The process will gain its own momentum.

Summary – The IDEEAL Decisions

You want the customer to make six decisions:

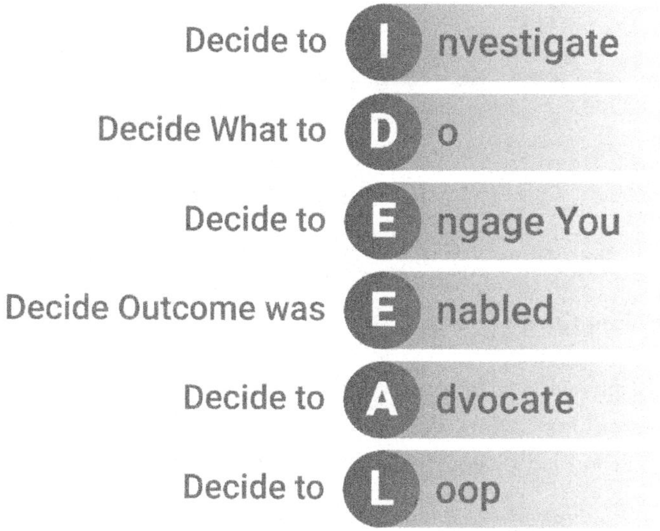

Figure 40: The IDEEAL Decisions

If prospects and customers make these IDEEAL Decisions, you will both prosper.

Section 2 – The Six Emotions You Need Customers to Experience

Through your outcome program, you will seek to have your customers make the six IDEEAL Decisions:

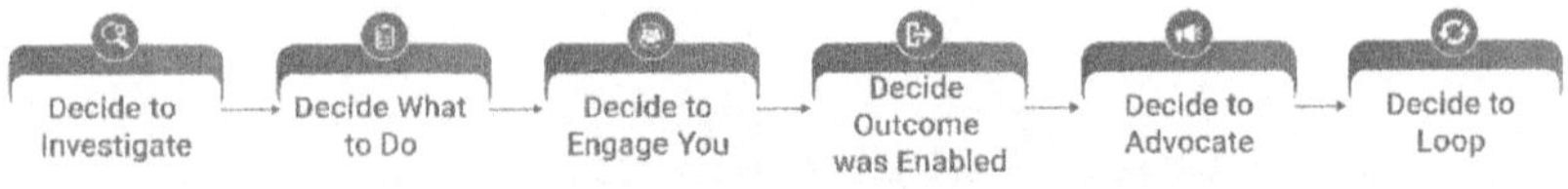

Figure 41: The six IDEEAL Decisions

How do you get customers to make these IDEEAL Decisions?

The Role of Emotions

A decision begins with an emotional commitment to a path of action. It then involves rationalisation. People will come up with the rationale to justify the decision led by emotion.

In the *Annual Review of Psychology* article 'Emotion and Decision Making' (Lerner, S. et al., 2015), the authors said: 'A revolution in the science of emotion has emerged in recent decades, with the potential to create a paradigm shift in decision theories. The research reveals that emotions constitute potent, pervasive, predictable, sometimes harmful and sometimes beneficial drivers of decision making.'

In this section we will explore the emotions needed for customers to make the six IDEEAL Decisions.

Decide to Investigate

Before any actions start, you need the Decider to consider their business results. Something must cause them to take time out of their busy day to ask: 'Are the results OK or do they need to improve?'

The Decider needs to feel curiosity.

Even then, if the results do need to improve, the Decider might not act. Inertia might win and they decide to do nothing.

The Decider needs to feel another emotion.

Seth Godin, one of the world's best-known marketing experts, observed: 'Awareness is important, but it is insufficient. Action comes from tension, desire and fear. Action is the hard part.'

(Godin, S., 2019)

A Decider might be aware their results should improve but it will take tension, desire or fear to break the inertia.

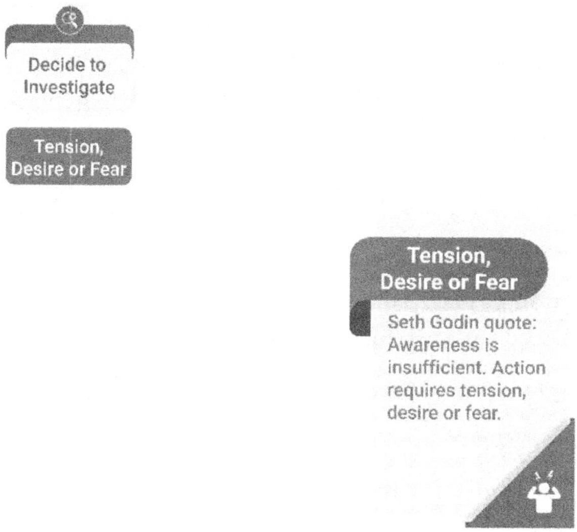

Figure 42: Tension, desire or fear

Decide What to Do

The Decider is now in the What to Do phase—they need a plan for a business initiative. By the end of this phase, they need to experience another emotion: confidence.

Analyst firm Gartner published an article in September 2019 titled 'The Secret to B2B Account Growth' (Starita, 2019). They found an interesting relationship between expansion revenue and confidence: 'When B2B buyers feel confident about their ability to navigate a new decision, they are 2.6 times more likely, on average, to expand an existing relationship than those who lack confidence.'

The quality of the plan for the business initiative will drive the Decider's ability to make the next decision. Putting forward a plan for a business initiative involves risk for the Decider. If it goes wrong, it will damage their reputation. Or worse.

They need strong confidence in the plan.

Another Emotion

There's another emotion in play. You want them to listen to you and follow your advice. You want to influence or shape the business initiative.

For this the customer needs to believe:

- They can trust you.
- You know what you're talking about.
- You're focused on what's important to them.

They need to develop faith in you. It may not yet be complete faith. But from the earliest of engagements, you need to work on building their faith in you.

You start this step with a focus on building faith in you. And you end with having helped them define a clear business initiative—you have helped them Decide What to Do.

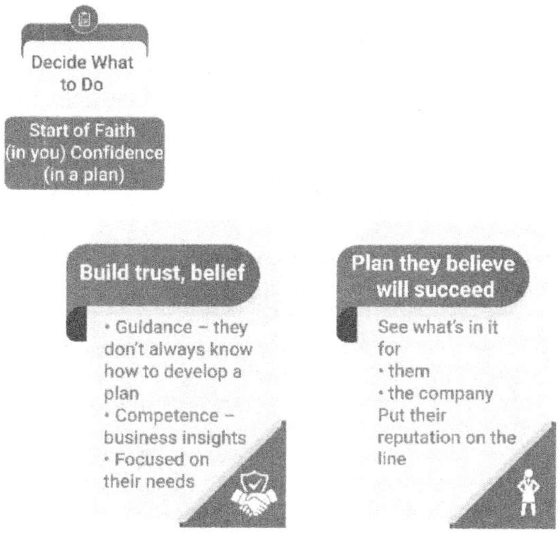

Figure 43: The start of faith (in you) and confidence (in a plan)

Decide to Engage You

The customer or prospect needs to feel good about working with you. You have convinced them to focus on the Success Outcome. You've helped them understand the PSPSPS they need to get right.

They want to believe you've aligned their interests with yours—which is true. If you jointly create measurable improvements in the Success Outcome, you will both reap the rewards.

They want to trust you will behave well in their engagement with you.

They want to feel you will work on their behalf. That you'll get the job done. And make them successful.

They want to have faith in you and your company.

To win the deal, you need prospects and customers to feel faith in you.

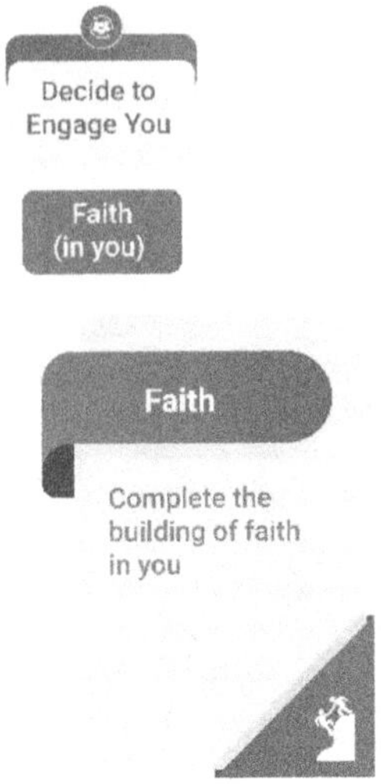

Figure 44: Faith (in you)

Decide the Outcome was Enabled

When a Decider moves forward with a business initiative, they take a risk. If it goes wrong, their reputation and career will suffer.

When they choose a vendor, they take a risk. If the vendor doesn't deliver on their promises, the Decider will suffer.

The decision to focus on a measurable business outcome also has risk. If it fails, again the Decider will suffer.

The Decider may have extra pressure. Perhaps people in the company denigrated the initiative, or the approval process required extra commitment.

At each step, the Decider takes a risk.

So when the business initiative succeeds, the Decider should feel one emotion: Victorious! They took the risks and it paid off. They need to feel it was all worth it.

To proclaim victory, they need:

- Clear expectations and a measurable objective at the start of the project
- A baseline against which to measure results
- Measurement of the results

You must measure the results with the customer for the Decider to feel victorious and Decide the Outcome was Enabled.

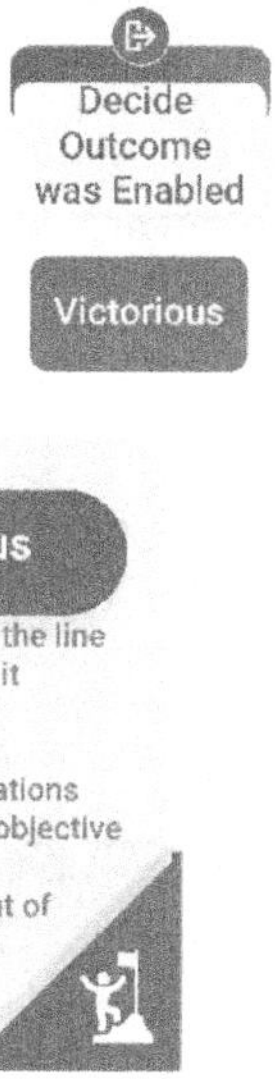

Figure 45: Victorious

Decide to Advocate

Victory does not automatically lead to advocacy. What if no one applauds their victory? What if higher management gives them a quick pat on the back but then moves on to other matters? The customer project team will feel deflated, maybe even resentful.

Feeling like this, they are unlikely to act as a reference for you—let alone become a proactive advocate.

You need their achievements to attract the recognition and reward they deserve—a bonus, a promotion or even just acclaim.

Your Decider and key team members need to feel hailed as heroes and heroines to Decide to Advocate.

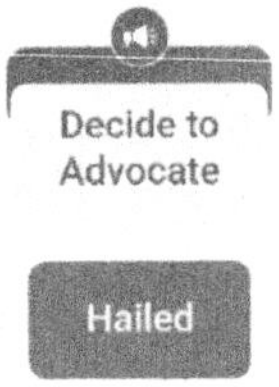

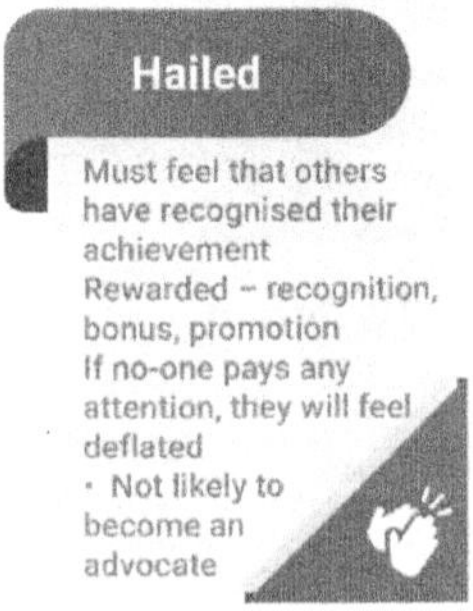

Figure 46: Hailed

Decide to Loop

The Decider and the team were victorious and have been hailed. Now you want them to commit to a repeating cycle of engagement where they face the same risks again.

There's one key reason they would agree. They think they can achieve victory and be hailed as heroes and heroines again.

You want them to anticipate being hailed again. You want them thinking forward to the approbation they can receive when they deliver even more success.

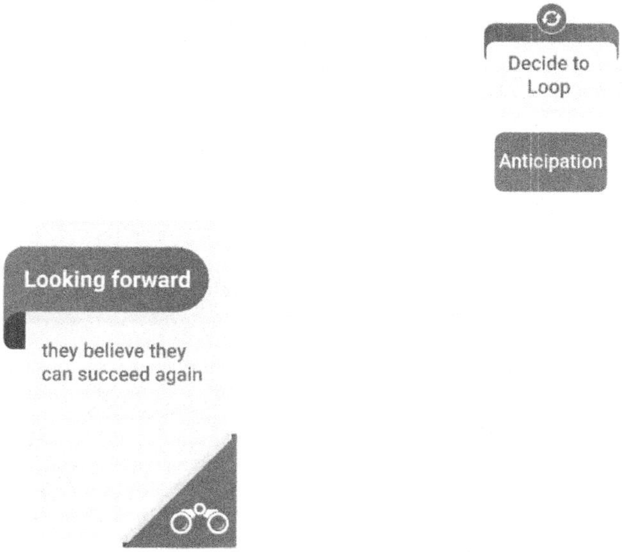

Figure 47: Anticipation

Strength of Emotion

For each of the six IDEEAL Decisions, the customer needs to feel emotion or sentiment to decide to move forward.

But emotion or sentiment isn't binary. It's not on or off; it has shades of grey. And there's a threshold of intensity needed. If the customer or prospect feels the emotion but it's weak, they will not proceed.

The level of emotion required to create action is called the Action Threshold. It's different for each of the six decisions you want customers to make—and it is also different for different people. But in each case the person must reach a threshold to overcome inertia. It looks like this:

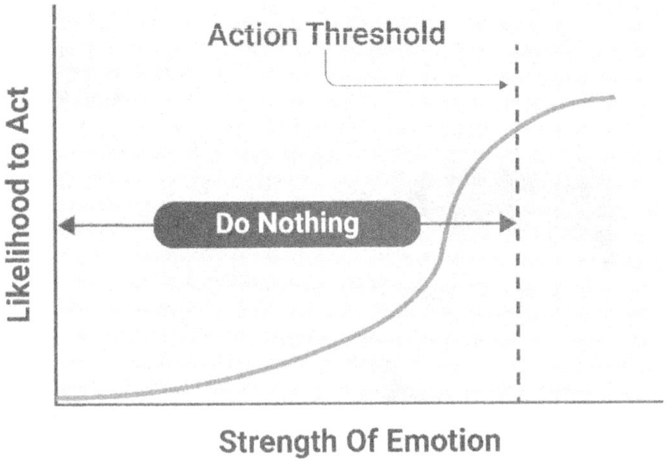

Figure 48: Emotions must reach the Action Threshold

Your outcome program must not only create the right emotions at the right time. It must create emotions strong enough to prompt the Decider to move to the next phase.

Summary

Your customers and prospects go on a journey, during which you want them to make six IDEEAL Decisions. To do that, they need to feel the right emotions at each step:

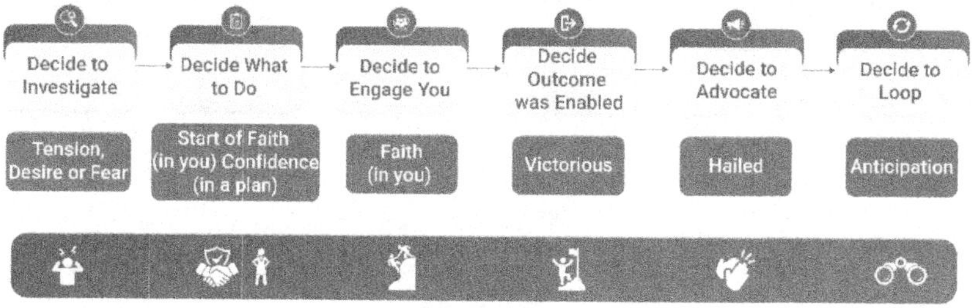

Figure 49: The six decisions and six emotions

At each step, the Decider must reach the Action Threshold for the required emotion to move on to the next phase.

Section 3 –The Six Steps of the IDEEAL Outcome Lifecycle

You know the six IDEEAL Decisions you want prospects and customers to make, along with the emotions they need to feel at each decision-point.

And you can determine the outcomes you enable for customers, your Outcome Structure:

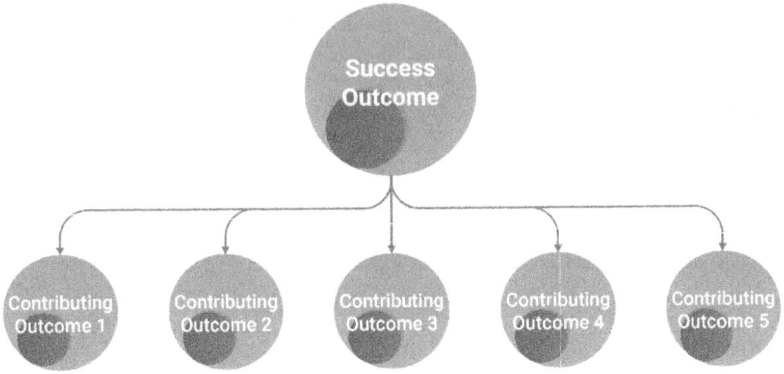

Figure 50: Outcome Structure

We'll now cover how to use your Outcome Structure to elicit the six IDEEAL Decisions.

The Six Steps of an IDEEAL Lifecycle

To grow revenue, your Outcome Lifecycle has six steps. It's a cycle that should repeat at least every year. Each step will drive one of the IDEEAL Decisions you want customers to make.

Figure 51: IDEEAL Lifecycle

These six steps in the lifecycle are also IDEEAL:

Figure 52: IDEEAL Decisions

You'll design your engagement differently for each segment. But the six steps will remain the same, as will the six decisions you want customers to make.

Investigation Starts

You'll design an approach to first create curiosity for the Decider. If they haven't already decided to investigate, you'll provide information to make them question their current results.

You'll then provide further information or insights (such as benchmarking) to create tension, desire or fear.

What to Do Defined

The customer or prospect now needs a plan to achieve the better business results. In B2B, customers and prospects spend money via a business initiative. Your engagement in this step should build faith in you and momentum for a business initiative.

This process will inform the Decider on how to achieve the better business results. And it will create the confidence needed for the second IDEEAL decision: Decide What to Do.

Engagement Won

Winning the engagement requires three steps. First, the customer must select you as part of their business initiative. You need the customer's selection criteria to focus on everything required to achieve the business result. Don't let it become a feature/function shootout.

Second, you'll need to agree contract terms.

Third, the initiative will need final approval before proceeding. You'll need to ensure the submission has everything required to achieve approval—the first time it is submitted.

During the What to Do Defined step, you'll create confidence in a plan you jointly created with the customer. As a result, the customer already has a strong feeling of faith in you.

During the Engagement Won phase, you complete that sense of faith needed for the third IDEAL decision: Decide to Engage You.

Outcome Enabled, Measured

The project to deliver the business initiative should have clear, measurable objectives. This is not an IT project, nor a vendor project. The scope includes everything required (PSPSPS) to deliver the business results. You will lead some PSPSPS elements, and the customer will lead others. It must run as a joint project.

You'll jointly focus on delivering the measurable business results from the project. You'll measure the results achieved and report the results back to the sponsoring customer executives. Your Decider will feel victorious.

Advocates Created

When you achieve a great project result with the customer,

you will ensure your Decider and key team members get credit. They will become heroes and heroines who will become references and perhaps even advocates for you. They will also help ensure renewal occurs and will sponsor you when it's time to find the next business initiative.

In this step, you're ensuring the key customer people feel hailed. That's what they need to make the fourth customer decision: Decide to Advocate. They will tell people how great you are at enabling the Success Outcome.

Cycle Looped

You will agree to a repeating cycle of engagement with the customer. It will emphasise continuous improvement of the Success Outcome. The next round of the cycle will begin with measurement of the Success Outcome and Contributing Outcomes. You'll formalise a process of reviewing the measures and identifying the next round of improvements. This process begins the Outcome Lifecycle again.

In this step, you're laying the foundation for your Decider and key people to be heroes and heroines time and again. That will create anticipation, which is needed for the fifth customer decision: Decide to Loop.

The IDEEAL Lifecycle for Your Larger Customers

To illustrate how the IDEEAL Lifecycle works, consider an approach for your larger customers. While you may choose a different approach, this example will give your insight into how an IDEEAL Lifecycle could work.

Investigation Starts

You begin with an Executive Business Review with each of your larger customers in which you:

- Discuss industry trends, the customer's competitors and new developments affecting the customer's business.
- Review metrics for the Success Outcome and Contributing Outcomes.

- Discuss ideas the customer executives have for improvements.
- Offer suggestions for improvement in the Success Outcome.

The information you provide, and the subsequent discussion, have one purpose. To create tension desire or fear among the executives.

The output should be a small number of possible improvement areas agreed by the executives as worthy of further analysis. In other words, the executives agree to investigate these possible improvement areas.

What to Do Defined

Next, you and the customer will jointly review each of the possible improvement areas from the Executive Business Review. At a high level, you'll determine if there's a business case for going ahead with each of the possible improvement areas. You'll jointly report back to the executives.

For those improvement areas with a strong business case, you'll help the customer create a detailed plan. The plan will include everything needed for the business initiative to gain internal approval, including a detailed business case.

This detailed plan should fill the executives with confidence.

Engagement Won

The customer must select you as part of their business initiative. On most occasions, the customer won't consider other vendors—you are integral to the planning of the business initiative. It would be high risk to bring in another vendor.

If they do consider other vendors, you want the selection criteria to focus on everything needed to achieve the outcome. And that's set out in the detailed plan, which has all the right headings. You can now help the customer turn those headings into selection criteria. This will stop the evaluation becoming a feature/function shootout.

When it's time for their final approval internally, the detailed plan has everything they need. This ensures a high-quality submission to the approving body.

You have continued your focus on the Success Outcome and now created the alignment needed for the customer or prospect to choose you.

Outcome Enabled, Measured

The project to deliver the improved business results should have clear, measurable objectives. The detailed plan has these project objectives, which you jointly developed with key customer staff.

You can now implement an outcome-based project approach. Achieving the business objectives requires more than just getting your products working—a joint project with joint accountability for the business objective makes the project a collaborative one.

Go-live is not the end of the project. You don't close the project until you have measured the results and reported back to the sponsoring executives.

Enabling a measurable business result lays the foundation for the rest of the process.

Advocates Created

When you achieve a great project result with the customer, it's time to turn your Decider and key people into heroes and heroines. Write stories or create videos about how well they did. Publish them internally and on social media.

Make it about your Decider and key people and they will become your advocates. They will Decide to Advocate.

Cycle Looped

You can now establish your repeating cycle of engagement with the customer. It will emphasise continuous improvement of the Success Outcome. Agree on how and when you will measure the Success Outcome, when Executive Business Reviews will take place and so on. In short, agree how you'll work together to drive further improvements in the Success Outcome. This process begins the IDEEAL Lifecycle again.

Your IDEEAL Lifecycle in Different Segments

We have set out the six steps for large customers. For other segments, you'll also adopt these six steps. What you're trying to achieve at each step will remain consistent. But the way you execute will differ.

For example, in both your large and small customers, nothing will happen until a Decider Decides to Investigate. In your large customers, you can prompt the Decider through a face-to-face Executive Business Review.

In your smaller customers, you can't afford this face-to-face engagement. But you may set up automated capture of business results and generation of a report for the Decider. In a smaller customer, the automated report may prove enough to prompt the decision to investigate.

Different execution, same outcome.

It's Always About the Success Outcome

The six steps of the IDEEAL Lifecycle will drive your outcome program. At every step, you focus on how to improve the Success Outcome for the customer.

Different business initiatives will have different objectives. But one thing will remain constant. Every business initiative will drive an improvement in the Success Outcome for the customer.

The Success Outcome remains the guiding light, the North Star for you and the customer.

Summary

We have now had a high-level look at the six steps of an IDEEAL Lifecycle.

Earlier, we discussed the six emotions needed to elicit the six IDEEAL Decisions.

Your engagement model now has a very simple concept underpinning it. You need:

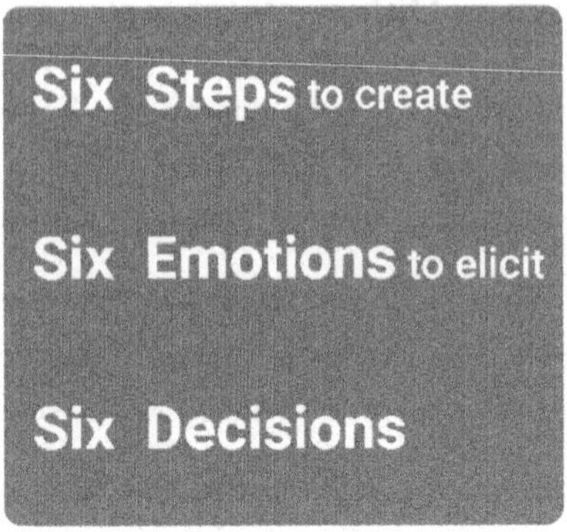

Figure 53: The Rule of Six

This captures the essence of How You Offer and Deliver. Now add What You Offer:

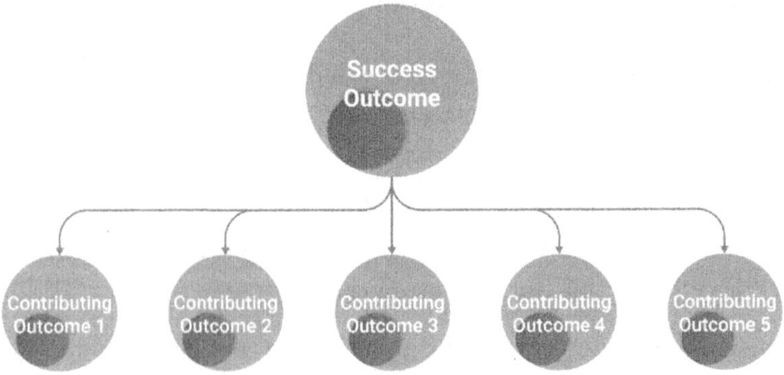

Figure 54: Outcome Structure

These two simple concepts underpin your entire outcome program—they will drive everything you do.

The rest of the book adds detail on how to make these concepts work.

But we still have lots of new ideas to introduce. For example, in Part 1 of Book 2, we will discuss why you should stop trying to build a pipeline of product opportunities. And what to do instead.

Splunk delivers business value by helping customers make sense of their massive data. David Caradonna, AVP Strategy & Programs for Industries & Specialization, shared his view:

Selling and delivering based on outcomes and value is established at Splunk through the Business Value Advisory practice. Launched in 2013, this practice helps customer deciders see the value of their data, through a Splunk lens. And we express this value in terms understood by a chief procurement officer or similar role.

Our outcomes and value approach covers the entire customer lifecycle. We begin by helping deciders see the 'future value' they can achieve through their relationship with Splunk. We also measure the 'value realized' from the outcomes we help each customer achieve at various stages during a customer's journey.

The Business Value Advisory practice helps at two levels. First, our value advisors have deep knowledge of what it takes to achieve a business outcome and create value. They work closely with account teams and customer deciders. They become trusted value advisers and they help our sales support engagements remain focused on delivering an outcome and creating value.

Second, the practice has developed an automated platform to help other Splunk staff help customers. It guides our staff on outcomes and value using outcome statements, value paths, use cases, ROI equations, product capabilities and maturity stages for each of our core markets. The staff simply need to facilitate value-driven conversations with their customers and follow the appropriate paths laid out for them to deliver outcomes and value. This allows us to scale and prescribe the right adoption plans across a growing number of customers and to help them realize the desired value and outcomes.

We adopt a collaborative approach with our customers. We jointly identify new ways to improve the customer's business. We jointly develop a plan for how to achieve the required outcome and value, and we then work together to enable the outcome and create the value.

This successful cycle typically leads to three outcomes:

- a loopback where customer deciders look for the next opportunity to improve their business
- internal advocacy in the customer for the collaborative approach with Splunk
- low friction renewals as realized value outweighs the investment

BOOK 2 –
MAKING THE CONCEPTS WORK

PART 1 – Outcome-based Lead Generation

Customers and prospects now engage vendors late in the buying cycle. This has created a void in engagement for vendors. In this Part, we'll discuss how this came about.

And then we will discuss how to overcome the void. In the early stages of the buying cycle, customers are building a plan for a business initiative. If you are to re-engage in this early stage, you will also need to work on the customer's business initiative. We'll introduce the BuILD process—Business Initiative Lead Development.

The BuILD process will allow you to re-engage early in the buying cycle. And it will set you up to win the deal.

Section 1 – The Void in Customer Engagement

Customers and prospects have changed the way they buy. They now engage vendors much later in the buying cycle. This late engagement has caused vendors' influence during the buying cycle to plummet. Vendors who want to re-establish influence early in the buying cycle need a new approach.

The Customer Buying Cycle

The buying cycle includes the first three decisions you need customers and prospects to make:

- Decide to Investigate—the Decider decides they need to improve results
- Decide What to Do—they develop a plan for how they will achieve the better results
- Decide to Engage You—they choose you to help them achieve the better business results

Figure 55: The first three decisions form the buying cycle

What Happened in the Past

In the early 2010s (and before), customers and prospects engaged vendors early in the What to Do phase. Vendors were a useful source of insight. They provided information on industry trends, what other companies were doing and the latest in technology. The dotted line indicates when customers/prospects previously engaged vendors.

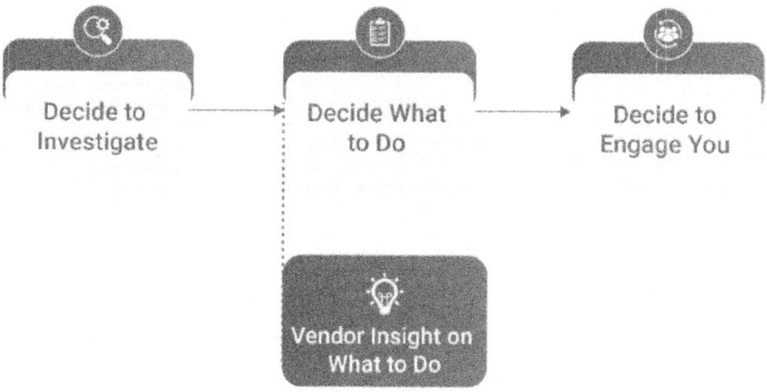

Figure 56: In the past, engagement with vendors occurred early in the buying cycle

Engagement in the What to Do phase provided vendors with two opportunities:

First, they could influence or shape the emerging business initiative. Vendors would emphasise problems they excelled at solving. And de-emphasise others.

Second, they could start to build faith. While engaging customers or prospects in their What to Do phase, vendors provided advice and guidance. Good advice and guidance resulted in a vendor increasing their faith level. And that increased the chance of making the sale.

And then Online Information Exploded

Today, customers and prospects can get the information they need online. Myriad sources provide insights into industry trends, what other companies are doing and the latest technology. Vendors' websites provide insight into their products and services. Online influencers have appeared and hold great prominence.

Customers and prospects no longer need to engage vendors in the What to Do phase. They can get the information they need elsewhere.

There is also a plethora of information about the vendors themselves, including their customers. This has pushed the engagement point even further out.

Engagement now commences well into the Decide to Engage You phase where customers are trying to decide which vendor to engage.

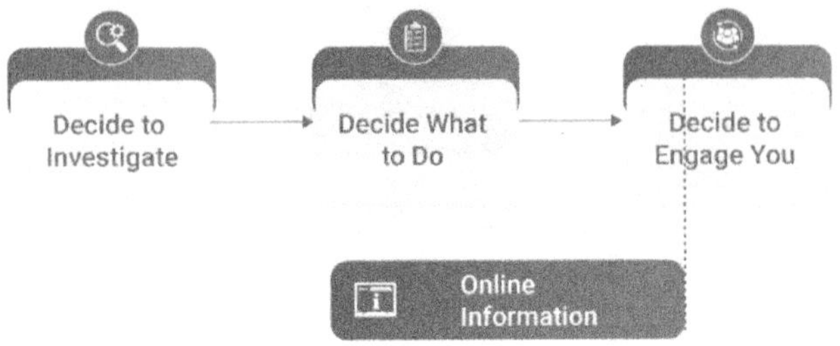

Figure 57: Today, engagement with vendors occurs late in the buying cycle

The Void

For vendors, the shift in engagement point has created a void.

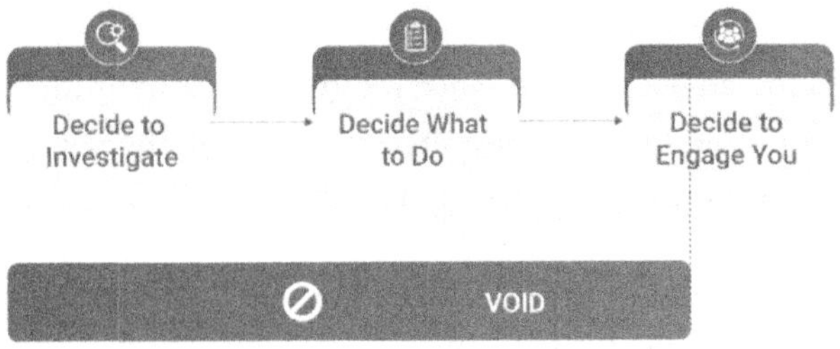

Figure 58: Late engagement has created a void in customer engagement

Vendors have tried to overcome the void. Most have a content strategy to attract attention in those early phases. But content won't shape a business initiative the way face-to-face engagement once did. Nor will it build the same faith.

Overcoming this void requires a new engagement model.

Section 2 – Overcoming the Void

Phase-appropriate Engagement

For each of the three decisions of the buying cycle, the topics a vendor talks about must be appropriate and relevant to the decision.

Imagine you're thinking about buying a new car. There's a process you go through.

First, you must decide whether you really want a new car. Something has prompted you to consider cars. Perhaps you see a new model and feel left behind. Or you have another child and need more space. Or you just feel like a change. If you feel enough tension, desire or fear, you will Decide to Investigate.

Second, you research. You look at various options. Whether formally or informally, you make a shortlist. You find out more about each vehicle on that shortlist to come up with a preferred vehicle. When you finish the research, if you feel confident enough, you'll decide to move forward. You'll Decide What to Do.

Finally, you will decide on a dealer to buy from (Who to Engage).

Imagine you haven't yet decided to investigate, but you are curious. So, you visit a showroom to get a sense of the vehicles. A salesperson approaches: 'Good morning, welcome to our showroom. Let me tell you about our great financing options.'

This would throw you. Financing is a topic that belongs in the third decision. But you still haven't made the first decision. You haven't yet decided to buy a car, let alone have any interest in financing.

Figure 59: Phase-appropriate engagement

Because the topic of financing isn't aligned with the phase you're in, it's not relevant right now. So, you tune the salesperson out, walk around the showroom for a while and then leave. That salesperson won't make a sale to you.

The salesperson should have found out what decision you were trying to make. And talked about topics relevant to that decision. For example: 'Good morning and welcome, what brings you in here today?'

Your answer informs the salesperson you're in decision one. They now know their task—they must create enough tension, desire or fear so you decide to move forward.

What's the moral of this story? You need phase-appropriate engagement with customers and prospects. The big issue lies with knowing when to talk about your products—and when not to.

Filling the Customer Engagement Void

The void affects the Decide to Investigate and Decide What to Do phases. To re-engage in the void, you need to address topics relevant for those phases.

Discussion of your products belongs in the Decide to Engage You phase. If you talk about your products in the first two phases, the customer or prospect will tune you out and walk away. Even case studies of the benefits other companies have gotten from your products won't be enough. They may create some curiosity, and may even create some tension, desire or fear. But a case study won't create a plan that fills the customer with confidence.

In the Decide to Investigate phase, you need to provide information, tools or guides that will create tension, desire or fear.

In the What to Do phase, you need to help the customer or prospect develop a well-defined business initiative. They need a plan that will deliver the better business results they need. By improving the Success Outcome. And you need to help them develop a great business case.

During both these phases, you can't talk about your products.

Section 3 – Creating Opportunities

We noted earlier Seth Godin's observation suggesting awareness is insufficient. Action comes from tension, desire and fear. Action is the hardest part (Godin, S., 2019).

We also noted analyst firm Gartner's research article (Starita, 2019) which observes customer satisfaction has a strong correlation with retention. But it has a low correlation with expansion.

In other words, making a customer happy will not create expansion revenue—that requires a Decider to feel tension, desire or fear.

We've also discussed a Decider needing a plan of action they feel confident will succeed.

Gartner's article added: 'When B2B buyers feel confident about their ability to navigate a new decision, they are 2.6 times more likely, on average, to expand an existing relationship than those who lack confidence.'

We can draw three observations from these insights:

First, customer satisfaction remains important. If customers don't feel satisfied, you will lose them. But customer satisfaction isn't a growth strategy. Customers won't spend more with you if you only focus on making them happy.

Second, to start a new buying cycle, a Decider first must feel uncomfortable with their current results. You should, therefore, aim to make Deciders uncomfortable and create tension, desire or fear.

Third, you want to achieve the 2.6 times likelihood of expansion identified by Gartner. So, you need Deciders to feel confident to navigate a new decision. They need confidence in a plan to achieve the new business results.

Let's explore how the IDEEAL Lifecycle delivers tension, desire or fear followed by confidence for Deciders.

Section 4 – Business Initiative Pipeline

You want to overcome the void early in the buying cycle. Customers and prospects must want to involve you. They must feel you can help them, in a phase-appropriate way. You can't talk about your products.

Business Initiatives

When it's time for a customer or prospect to get final approval, the submission will describe a business initiative. It will set out how the business will change. And it will have a strong business case.

There's a clear output needed from the Decide to Investigate and Decide What to Do phases. The customer or prospect needs a plan for a business initiative. It needs to be well-scoped, clear and must fill the customer or prospect with confidence. They will then go on to the Decide to Engage You phase.

You need to ensure customers and prospects develop a plan for a business initiative. That's your focus in lead generation.

You aren't going to try to build a pipeline of product/service opportunities. You're going to build a pipeline of business initiatives.

When you do this well, the customers or prospects will move forward to the Decide to Engage You phase. You will build their faith in you through helping them develop the business initiative.

And you will know the business initiative better than any other vendor could hope to achieve. You are the low-risk choice for the customer or prospect—the natural choice. You have an unfair advantage over the competition.

While you are not guaranteed to win, the deal is yours to lose.

Section 5 – Business Initiative Lead Development – the BuILD Process

You need to build a pipeline of business initiatives. You'll run a formal program to achieve this pipeline. You'll use the BuILD process: Business Initiative Lead Development. It addresses the first two decisions you want customers to make.

In the Decide to Investigate phase, you'll create curiosity, then tension, desire and fear.

In the Decide What to Do phase, the BuILD process has two elements:

- Assess—the customer will consider the high-level options for achieving better business results. And then decide on a high-level path to success
- Plan—the customer will develop a detailed plan for the business initiative they want to pursue

The BuILD process, and where it fits, looks like this:

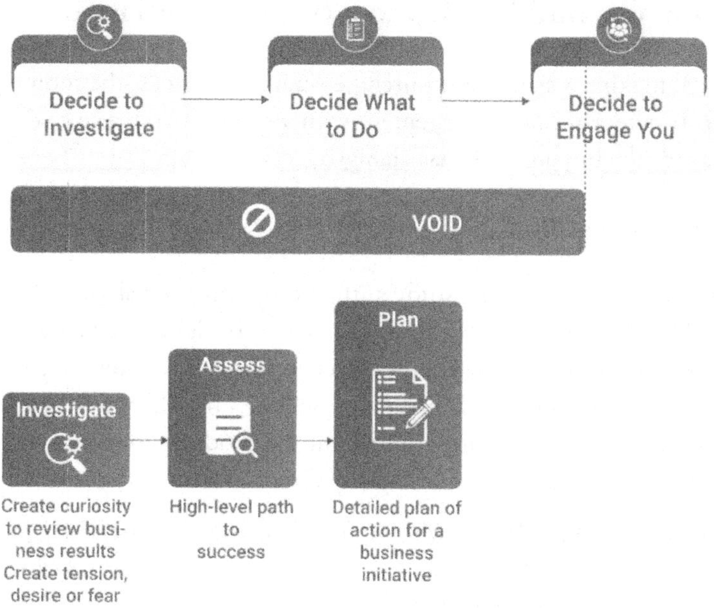

Figure 59: BuILD – Business Initiative Lead Development

Section 6 – Building Up to a Business Initiative

In B2B, making a significant purchase is rarely a spur-of-the moment event. It takes time and engagement with a vendor. And, as we've discussed, the buying cycle has changed.

Zero Moment of Truth

In 2011, Google released a study authored by Jim Lecinski called the 'Zero Moment of Truth' (Lecinski, J., 2011). It highlighted a change in buying behaviour. It showed how buyers researched a potential purchase online before making any contact with sellers. The sellers were no longer the primary source of information about the products they sold. Buyers were well informed before making any direct contact with the sellers.

The report also highlighted what it takes before buyers are ready to purchase. For significant purchases, people don't make snap decisions. They spend time and effort before they are ready. The time and effort inform them. It also helps them feel confident with the seller. And with their own decision to purchase.

While the 'Zero Moment of Truth' study focused on consumer behaviour, it has strong lessons for the B2B world. After all, even in B2B, it's still people who buy.

Time and Engagement

Significant purchases require a combination of time and different types of engagement from the buyer. For a seller to increase the chance of making a sale, you need to offer:

- Multiple types of engagement: for example, email, snail-mail, phone calls, texts, videoconference, face-to-face meetings, instant messaging, mobile apps such as WhatsApp, your website and content marketing
- Lots of engagement time: enough for them to build faith in you
- Lots of touch points or separate engagements

Building Engagement

You'll want to utilise a series of small incremental steps in commitment by the customer or prospect. At each step, the customer commits a little more time, effort and eventually money.

The first step should require minimal commitment. It's probably free. It should aim to convince the customer or prospect to take the second step.

The second step will require the customer to spend a bit more time and effort. It should aim to convince the customer to take the third step.

The third step will take the customer or prospect more time and effort than previous steps. And convince the customer to move to the next step.

A series of steps with small increases in time, effort and eventually money underpins the BuILD process.

Section 7 – BuILD Process – Investigate

In Step 1 of your BuILD process, you aim to have a Decider Decide to Investigate. They need to experience tension, desire or fear. And it needs to be strong enough to overcome inertia.

You won't engage with all prospects and customers the same way to create that tension, desire or fear.

You can divide customer or prospect spend into two major categories:

- 'Involuntary' spend—a customer or prospect has little choice but to spend money with someone
- Discretionary spend—requires good reasons for them to spend with anyone

Before the start of a buying cycle, Deciders can exist in one of two states:

- Decided to investigate—the Decider has already decided results need to improve and they should do something
- Not yet decided to investigate—the Decider may be unaware of the need to act. Or may be aware but has not yet decided

You can use the two types of spend and the two states to create a simple matrix:

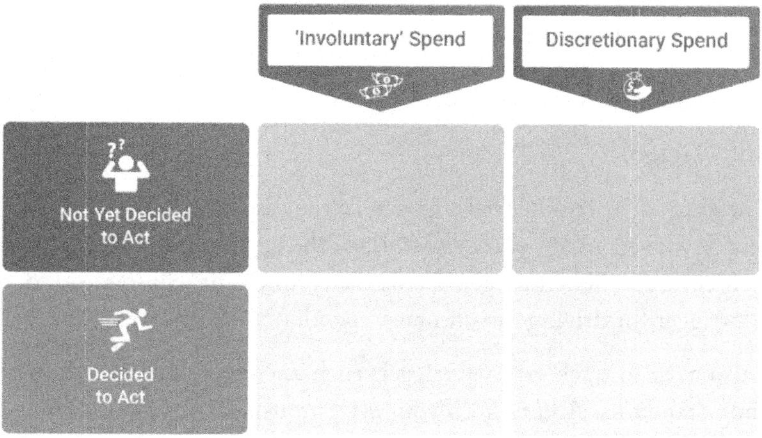

Figure 60: Two types of spend and two states of the Decider

Involuntary Spend

We place the word involuntary in quotes as it's rare for a customer or prospect to have no choice at all. They can usually find ways to achieve a result without spending money. But sometimes spending money provides the most practical approach. They won't always be happy to spend that money, but they do.

Examples of 'involuntary' spend include:

- Growth of the customer's business resulting in an increased number of seats or subscriptions required. They must buy more seats or subscriptions (existing customers).
- Changes in regulations or other compliance requirements—for example, changes in accounting standards may require companies to buy new products. Or, for companies in regulated industries, new regulations may require additional investment. The customer may not feel there is much value in their spend but have little choice but to go ahead.
- Mergers and acquisitions—a company acquiring another may have standard software they use. The acquiring company may

insist the acquired company use the same software. The acquired company has little choice but to buy the new software.

With 'involuntary' spend, you'll adopt two steps.

Identify Triggers

You need to identify potential triggers. In the case of existing customers and usage management, you need to track the customer's usage compared to their contract. When usage exceeds the contract, or will soon do so, you have a trigger for driving 'involuntary' spend.

You also need to track external triggers such as changes in regulations or business standards. And to keep your eye on business activities such as mergers and acquisitions. These can all be triggers.

Contact Deciders

Having identified a trigger, you will contact Deciders. You need to make sure they feel uncomfortable, i.e. they need to act. In some cases, the Decider will already have decided to investigate—they were aware of the trigger and knew they had to do something.

Some Deciders may not know of the trigger. In this case, you can help them understand why they need to investigate. Other Deciders may know of the trigger but have not yet reached a conclusion on whether to investigate. Again, you need to help them feel uncomfortable enough that they Decide to Investigate.

For 'involuntary' spend, our matrix now looks like this:

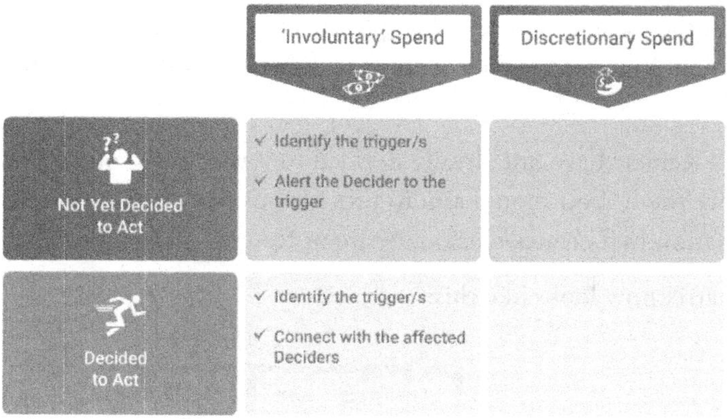

Figure 61: 'Involuntary' spend

Discretionary Spend

The need for business change drives discretionary spend. It's always about business change. Your outcome program will have its strongest effect with discretionary spend.

Some possible catalysts that may create the necessary tension, desire and/or fear for a Decider include:

Catalyst	Examples
Budget Increases	The Decider must decide if business change is needed to achieve the higher budget numbers.
Competitor's New Activity	The Decider will worry that the competitors provide a new threat and will need to decide whether to respond.
New Processes or Technology	Is the new technology a threat or a boon? The decider needs to reach a conclusion on whether to do something.
New Perspective on Their Business	The Decider learns of a possible new to-be state, something not previously considered. Should they act?

Figure 62: Catalysts for Discretionary Spend

When Deciders have already decided to investigate, they have already been through the tension, desire or fear phase. They know they need to do something. You need to find these Deciders early in their What to Do phase.

When Deciders have not already decided to investigate, you will need to convince them. You should aim to provide business insight. Give them information that creates tension, desire or fear.

The matrix now looks like this:

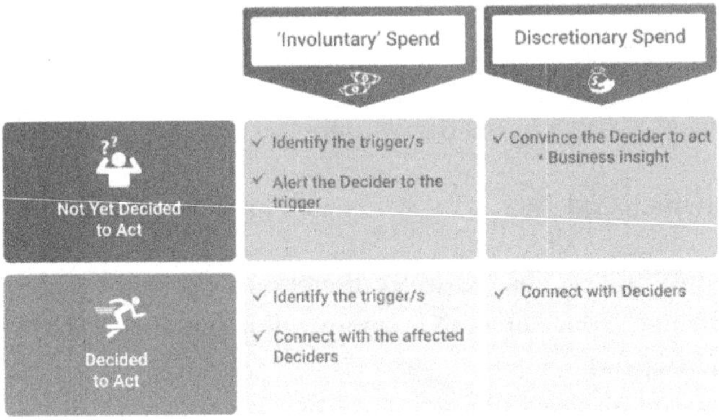

Figure 63: Discretionary Spend

How you execute will vary depending on segment. But what you're trying to do remains consistent across segments.

Summary

Triggers drive 'involuntary' spend. You need to identify the triggers and then connect with the affected Deciders.

Business change drives discretionary spend. If the Decider has already decided to investigate, you don't need to sell the business change. If the Decider hasn't decided to investigate, you need to provide insight. You need them to feel tension, desire or fear to catalyse a decision to investigate.

Section 8 – BuILD Process – Assess

Customers and Prospects Don't Always Know How to Proceed

Often, customers won't have a clear idea of how to develop the plan for a business initiative. You can act as a guide.

Passive engagement, such as providing a white paper or case studies, may help Deciders Decide to Investigate. But it's unlikely to help them Decide What to Do.

You want to engage actively. You can engage face to face or electronically but you need to actively guide them.

If we add this concept, the matrix now looks like this:

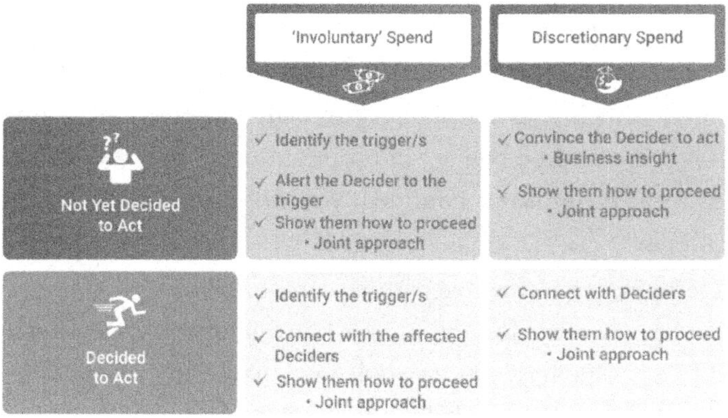

Figure 64: BuILD process – Assess

Building Faith in You

In the Decide What to Do phase, you begin the process of building the customer's faith in you. People will not buy from people or companies in whom they don't have faith. Nor will they listen to people in whom they don't have faith.

Your BuILD process presents a wonderful opportunity to build faith. In the BuILD process, you focus on the business results, the outcome, the customer needs and, as a result, build faith in you.

What Is Faith?

In simple terms, we have faith in a person or company we believe will act well towards us. Some ideas that might influence people to believe you will act well towards them are:

- Consistency of experience over time—you behave in a consistent and predictable manner.
- Most of the truth—customers must believe what you tell them is largely true. Customers aren't naïve enough to believe they will get 100% transparency (who would believe that?). But they need to feel confident you will not misrepresent facts nor leave anything significant out.
- Fairness—they must believe you will act to balance your company's interests and their company's interests in a fair way. They understand you are in business, just as they are. But they must feel completely confident you will never act against their interests.
- Competency—this covers three components:
    - Knowledge—you must know your subject matter
    - Gravitas—your people must have the personal presence for others to believe their advice and be willing to follow it
    - Insight—you show you can teach them things they didn't know about your subject matter.

In a Salesforce survey, 80% of customers felt their experiences (based on trust and understanding) were as important as the goods and services (Salesforce, 2018).

Building faith matters.

Assess – High Level Only

The first step you ask the Decider to take in developing a plan for a business initiative needs to feel like a small step. If you suggest a process requiring a big commitment of time and effort, the Decider may baulk. And do nothing.

So, they need a high-level plan for how they will improve their business results. Help them develop that quickly.

In some customers, you will work directly with the Decider on this step. In others, the Decider will delegate the analysis and development of a high-level plan.

Either way, you need a plan that includes:

- A clear path forward—you and the customer agree how to achieve the better business results. This is a high-level description, not a detailed plan.
- A high-level assessment of the business case. The Decider will need confidence it makes sense to move forward with this potential business initiative.

It's rare for a Decider to move forward with a business initiative with no personal win in mind. While you assist in the development of this high-level plan, you should seek to understand the Decider's personal win. If the high-level plan reinforces the personal win, you increase the chance of an initiative moving forward.

Confidence

The high-level plan will increase the Decider's confidence in their decision to act. You have shown there is a positive business case. And you've shown the customer how they might achieve the better business results. They are feeling good.

But unless it's a very small business initiative, the Decider won't yet approve the initiative. For any significant initiative, the Decider will need more detail.

Section 9 – BuILD Process – Plan

Detailed Business Plan and Business Case

Now, you'll help the customer develop a detailed plan for the business initiative.

You will again ask the customer to spend more time, effort and possibly money. Again, it will be an increase over that required for the high-level plan. And again, it won't be too big an increase.

You will now help the customer create a report containing everything required to achieve approval for the business initiative.

In face-to-face customers, you will seek their CFO's guidance on the format required for approval of a business initiative.

Armed with this insight, you will write the report in the format required for submission.

Joint Report

It's key that you work with the customer to prepare this report. It should not be an analysis and report by you alone. Key people from the customer will work with key people from your organisation to create the detailed report together.

The report will cover everything required to achieve the better business results, including processes, systems, people skills, partners and supporting elements. This business initiative may affect all five PSPSPS elements.

The report will include:

- Description of the to-be state—how the business will run at the conclusion of the business initiative
- What will change—description of the changes that will occur in the business to achieve the new to-be state
- Business case—a detailed business case in the format required for approval

- Project plan—a detailed plan setting out the time and resources required for this business initiative

Paid Engagement

If you work face to face in developing the report, it will take a lot of effort. You should seek payment from the customer for your work.

You can offer two main reasons for seeking payment.

First, you will remain product agnostic. You know talking about products now is not phase-appropriate and will undermine their faith in you. So, you will remain product agnostic. Any references to products will remain generic. The report could be used with any vendor. It will contain no direct references to your products or services.

Second, the report will address everything required to achieve the better business results (PSPSPS).

Thus, you're not engaged in a sales activity—it is a true consulting engagement. And customers should expect to pay for consulting.

We call this paid engagement a 10% Sale (more details in Part 6, Section 2 of Book 2).

Confidence

The detailed plan now has everything needed to get approval for a business initiative. And that should give the Decider complete confidence to proceed.

Success Consultants

You will execute your BuILD process more effectively if you have Success Consultants. Success Consultants understand everything the customer requires to achieve the Success Outcome. They guide the customer in developing the high-level plan. And will lead the work on the detailed report (more detail in Part 6, Section 1 of Book 2).

Profitwell delivers subscription revenue increases, automatically. Patrick Campbell, CEO, said:

We begin a potential relationship by helping a decision-maker see they could be doing better. Having gotten the decision-maker's attention, we first make a free service available. Profitwell Metrics sets up quickly, and immediately provides the decision-maker with insights into their subscription business. For free, forever.

This free service does two things. First, it helps the decision-maker identify where they could be doing better. It helps them develop a high-level plan on how to get better results. Second, it builds faith in Profitwell. The decision-maker gets real insights, and as promised, it's free.

Once they decide to improve their results, we have the tools to help them. And we have a clear pricing policy. The customer pays for results. They pay only when they reduce churn, or improve revenue recognition, or increase pricing effectiveness.

And that pricing policy also does two things. It gives the decision-maker confidence to proceed with a business initiative. And it completes the building of trust and faith in Profitwell. The decision-maker is comfortable to move forward.

Section 10 – Who Does What?

Who owns and runs the BuILD process?

There's no right answer—different vendors will make different choices.

Larger Customers

You may choose to engage larger customers face to face. There's enough revenue to justify the cost. You may engage physically or remotely. Either way you will talk with them one-on-one.

Traditionally, account managers own this segment and jealously guard their 'patch'. They decide what happens with their customers.

But as we noted in the Un-Problem Unexpanded, account managers are not renowned as systemic or disciplined.

But an expansion program needs a systemic, disciplined approach. You want a repeating cycle of engagement. You will jointly focus on continuous improvement of the Success Outcome. And you'll repeat the cycle every year (at least).

Traditional account managers may struggle to lead this model.

In many B2B companies, a new group called Customer Success (CS) should drive expansion (upsell and cross-sell) in the customer base. But most CS professionals don't want to be 'product salespeople'.

A BuILD process offers a great solution. Customer Success can set up and run the repeating cycle of engagement with customers. CS can take the customers through the Decide to Investigate and Decide What to Do phases. They can build the pipeline of business initiatives that will drive your expansion revenue.

CS will create revenue but will not be 'selling the product'. They will be helping customers build plans for business initiatives.

So, you might choose to have CS lead the systemic, disciplined coverage of these customers.

When a business initiative enters the Decide to Engage You phase, you can then decide the role of the account manager.

Smaller Customers

You'll have customers who don't create enough revenue to allow face-to-face engagement. You'll set up a self-help model to cover them.

Your self-help engagement model design will need the combined skills of several departments. But marketing might prove the most effective for the day-to-day running.

New Business (New Logo) Prospects

Engaging prospects will begin as a self-help model. As such, marketing may again prove the best group to run this aspect of BuILD. When you reach the Plan step, you may choose to engage large prospects face to face. For smaller prospects, a continuation of the self-help model will work.

Section 11 – How BuILD Differs from Traditional Lead Generation

BuILD Overcomes the Void

We've discussed the need for phase-appropriate engagement. BuILD helps you engage early in the buying cycle. It helps you engage jointly with customers to identify and gain approval for business initiatives. Initiatives that will need your products and services.

BuILD Defeats Present-Dayism

Solution selling and similar programs focus on identifying current customer problems. Then showing how your products provide a solution.

The challenge for customers lies with the pace of change in business. The current process takes time. Customers must:

- Agree with the vendor on the current problems to address.
- Develop a plan for a business initiative.
- Choose a vendor.
- Implement the solution.
- Wait for the settling-in period to complete.

By the time this has completed, a whole new set of problems has arisen. The customer never gets ahead.

An outcome program offers a different focus. A Success Outcome defines a new to-be state for the customer. It's an ongoing business result higher management of the customer regards as a success. Achieving the to-be state solves current problems. BuILD borrows from the wisdom of Stephen R Covey and his *Seven Habits of Highly Effective People* (Covey, 1988)—start with the end in mind.

Your outcome program addresses the problem of present-dayism. And it starts with the BuILD process.

Closer Relationship With Customers

Solution-selling approaches emphasise how your products or services will solve problems. But we have already discussed that achieving better business results requires PSPSPS. For your customer or prospect to achieve success, they need to get all five elements of PSPSPS right.

An outcome program, led by your BuILD process, embraces PSPSPS from the start. It's holistic. The need to get all aspects of PSPSPS right means both parties must collaborate. It becomes a joint project, with shared responsibility and accountability.

And that results in a closer relationship with customers.

Unique Benefits

We've explained the financial, market positioning and internal benefits of an outcome program. While the BuILD process contributes to all the benefits, three of them come solely from the BuILD process:

Expansion

You will develop a repeating cycle of engagement with your customers. You will jointly focus on continuous improvement of the Success Outcome. With each engagement, you will identify a business initiative to improve the Success Outcome. These business initiatives will drive your expansion revenue.

New Business Win-rate

You will improve your win-rate for three reasons. First, you'll stand out from competitors. Your Success Outcome will create a mental picture higher management in a prospect will want to achieve. And you will be the experts at enabling this outcome.

Second, you'll engage decision-makers (Deciders). Your ability to provide insight on a successful business outcome will get you to the executive level—and keep you there.

Third, you'll create more references and advocates. When those key customer people change companies, they will bring you into their new company.

Medium-term Pipeline

Your focus on continuous improvement of the Success Outcome will drive your expansion program. It will identify short-term opportunities. But it will also identify business initiatives the customer isn't yet ready to start. They are opportunities for next year. Across your customer base, this will create a medium-term pipeline.

Summary

The availability of online information changed the buying cycle for customers and prospects. It created an engagement void for vendors.

Your BuILD process will overcome this void. You'll provide insights in the Decide to Investigate phase that will create tension, desire or fear. And cause Deciders to Decide to Investigate.

You'll help them create a high-level plan to achieve better results. And then help them build a detailed plan for a business initiative containing everything needed to get approval.

Guiding them on this process will build their faith in you.

Extensis helps creative teams control creative chaos. Toby Martin, CEO, said:

We introduce the idea of creative chaos to the head of a creative team or the owner of a creative agency (the Decider). It might be through content, webinars or direct engagement.

Creative chaos arises when creative teams struggle to store, organize, find, share or control digital assets such as pictures, videos or their creative work. It drags down productivity. And it affects every creative team. But most people don't think of it in those terms.

The term grabs the interest and curiosity of the Decider.

We then offer a high-level assessment of how creative chaos affects the Decider's creative team. We call this a Chaos Assessment. The Decider answers 15 questions. We can then offer high-level insight into where the creative chaos is under control, and where it's causing issues. We also do a high-level estimate of how much the chaos costs the Decider.

The Chaos Assessment Report creates tension, desire or fear for the Decider. They realise they could better control the creative chaos. And they now have a high-level insight into where they could make improvements.

The final stage involves helping them develop a plan for controlling the creative chaos. We call this a Chaos Study. The resulting Chaos Report gives the Decider everything they need to know to proceed with a business initiative.

Importantly, we don't talk about our products or services throughout this process. We focus on helping the Decider build a plan for a business initiative. And that helps them develop faith in us. Which gives us the winning edge we need to win the deal.

PART 2 – Selling Using Outcomes

Kate Johnson, Microsoft's US President for four years, said: 'Delivering business outcomes to customers is Microsoft's only sustainable competitive advantage' (Dove, M., 2021).

You've built a pipeline of business initiatives through the BuILD process. You now need to win the engagements. And go on to enable the business outcomes Kate Johnson refers to.

For now, we will set out the steps for working with an existing customer who has been through your BuILD process.

We will cover customers that haven't been through your BuILD process and New Business (New Logo) prospects later in the book.

In Part 2, we will discuss how you can influence the selection criteria in your favour. We'll discuss how your BuILD process has set you up to win the deal, what happens when the customer has not been through the BuILD process, and then discuss the final two steps in winning the deal.

Section 1 – Selection Criteria

Through the BuILD process, you've helped the customer develop a business initiative to improve business results. You've then helped them create a detailed plan for the business initiative. And you've addressed all five PSPSPS elements, because achieving a business outcome requires all five.

PSPSPS should underpin the customer or prospect's selection criteria. The customer should evaluate each possible vendor on the vendor's ability to improve the Success Outcome. If a vendor cannot assist the customer with all five elements of PSPSPS, the customer should score that vendor lower.

Detailed Plan

The detailed plan contains the selection criteria covering all elements of PSPSPS.

Go through the headings in the plan and help the customer extract the right ones to set up selection criteria that include all elements of PSPSPS.

The engagement model now looks like this:

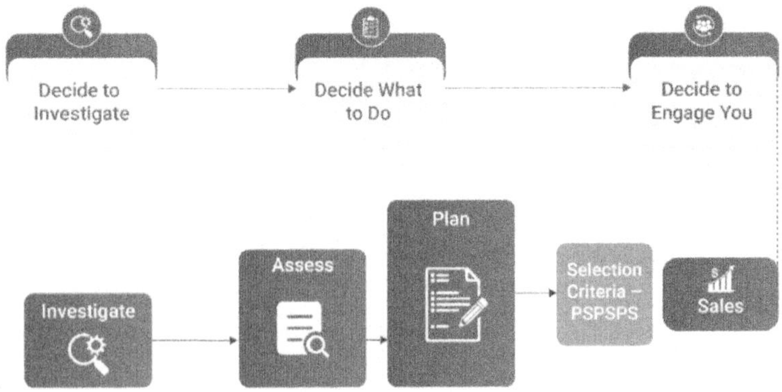

Figure 65: Influencing selection criteria

Section 2 – No Competitive Evaluation

Through each stage of your BuILD process, you've helped the customer. You have provided guidance and insights on everything required to achieve the better results they need. You know this business initiative as well as they do.

And at every step, you have remained product agnostic and focused on their business results.

You have built their faith in you. For the customer to introduce another vendor into this initiative would create high risk. Another vendor would not know the details of the initiative as you do.

It's likely, therefore, the customer will choose you. Without conducting a competitive evaluation.

Some customers may ask you to conduct a proof-of-concept. But over time, they may not feel the need for even this small step.

They have faith in you.

Competitive Situation

You will encounter customers who may not have a choice but to conduct a competitive evaluation. Some large organisations and government departments have policies requiring this approach.

You'll keep them focused on the selection criteria based on PSPSPS.

And that will give you the edge on competitors.

Section 3 – What If the Customer Plans a Business Initiative Without You?

Some customers may plan a business initiative without your involvement. You don't discover this until they are ready to select vendors.

The adoption of an outcome program is an evolution, not a revolution. It doesn't happen overnight. Sometimes, customers aren't aware of your outcome approach.

In this case, you disrupt their planned approach—in their own interests. You help them understand what it will take to achieve the business results they need. You introduce your Outcome Structure and PSPSPS. You talk about how you can help them with most of what's needed. And why they should not focus on the features and functions of products alone.

Then take them back to your BuILD process. Offer to help them develop a plan for everything required to achieve the business results.

If you're successful, the customer will change their selection criteria. And that will make winning much more likely.

Customers Who Don't Want an Outcome Approach

Some customers may know of your outcome approach but want to continue with a transactional engagement. They don't want your help to develop a plan of action.

You are now in a traditional sales shootout. Assuming they are a customer you want to keep, you have little choice but to adopt a traditional sales approach. If you win, it will likely involve much larger discounts than you'd like. But you don't want to lose customers.

This reinforces why you should move as many of your customers as possible to your outcome program.

Section 4 – Closing the Deal

After the customer has selected you, two activities remain:

1. Negotiating the Price

You and the customer must agree on commercial terms. Several factors will assist you in negotiating favourable terms.

Value of the Business Outcome vs Value of Your Products

Vendors will often attempt to sell the value of their products. Customers sometimes are loathe to agree. They know achieving the benefits will require their own efforts. So, they don't feel it's fair to assign those benefits to the vendor alone.

Under your BuILD process, you and the customer have identified the benefits arising from the initiative. You and the customer jointly documented the benefits. You don't need to 'negotiate' over the likely benefits. You know them.

The likely benefits now become the starting point for the negotiation of the price for your products.

Experts

We've discussed the idea that experts can charge more. Under your outcome program, you're no longer just a provider of great products and services. You're now an expert in enabling the Success Outcome. As an expert, you should expect to charge more.

Trust

People will pay more for products and services from companies they trust. Think about something as simple as pain relief medication. Lots of people pay a higher price for a brand they trust.

Through your BuILD process, you have developed faith and trust. The customer wants to work with you in preference to anyone else. Now, in the negotiation, you can benefit from that.

2. Obtaining Approval for the Initiative

Having negotiated commercial terms, the business initiative needs final approval. Sometimes the initiative will fall within the approval authority of a single executive. If not, the customer will need to make a submission to the board, an executive committee or some other approving body.

A poor-quality submission risks being rejected or put on hold. You need to ensure a high-quality submission.

The detailed plan you co-wrote in the BuILD process provides this high-quality submission. Before writing the report, you sought the advice of the customer's CFO on the format required. In writing the report, you included everything required for the submission. You ensured it had a great executive summary, a compelling statement of the benefits and a clear path for achieving the outcome.

Your report will increase the chance of the initiative gaining approval on its first submission.

Summary

Through the BuILD process, you helped the customer write a detailed plan for a business initiative. A plan that addresses all elements of PSPSPS.

You used this report to establish the selection criteria, which also focus on all elements of PSPSPS.

It's unlikely the customer will conduct a competitive evaluation. You know the business initiative as well as the customer does. And it would be risky for them to bring in a different vendor.

If a competitive evaluation does take place, you're well-situated to dominate and win.

Once selected, the detailed plan will help you negotiate a better deal. And help ensure the business initiative receives final approval.

PART 3 – Outcome Enabled, Measured

You've won the deal. The customer signed the contracts.

Now, you need to enable the outcome from the business initiative. When you do this, you create a track record of enabling business results. And that makes it easier to win later deals.

In Part 3, we will cover how to enable the business results using an outcome approach.

Section 1 – Step 4 in the IDEEAL Outcome Lifecycle – Outcome Enabled, Measured

A sale isn't the guarantee of profits it once was. Before subscription pricing, vendors received most of the money upfront. And that meant you would make a profit when you signed a deal.

So, sales teams felt getting the deal was all that mattered. And to a large extent that was true. Customers felt locked in. They spent so much early in the relationship that cancelling would result in a large write-off. And embarrassment for those executives who had sponsored the deal.

Subscription pricing changed the landscape. Vendors no longer receive most of the revenue upfront. It's spread over time. In the medium to long term, that's good for the vendor. Predictable recurring revenue makes planning, management and forecasting much easier. And it certainly helps stock prices. The market loves growth in subscription revenue.

But with every new deal won, there's a challenge for the vendor. Customers remain uncommitted. They contract for the smallest possible subscription. And they promise to buy more as their usage grows.

If the customer doesn't feel victorious, they won't grow their subscription. And they may reduce or even cancel it. Regardless of whose fault it is, if the customer doesn't feel successful, the vendor will lose future revenue.

Making the sale is just the first step in obtaining the full revenue from a deal. To bring in all potential revenue, the customer must feel victorious.

It's a Project

Customers should execute a business initiative as a project. If they do so, they are more likely to achieve success.

We need to show them how to run an outcome-based project.

A Different Mentality

Traditional thinking said:

- Sell your products.
- Get your products adopted.
- Later, try to sell more usage and more products (upsell and cross-sell).

An outcome approach says:

- Sell a business initiative to improve the Success Outcome.
- Deliver the results from this business initiative.
- Later, look for the next way to improve the Success Outcome, creating the next business initiative (continuous improvement of the Success Outcome).

Business initiatives should be something the customer can digest. The first time, try a minimum-viable outcome approach. Get a win.

You should reinforce that a business initiative isn't a one-off project. It's the beginning of a continuous improvement process. You don't have to enable everything in the first attempt.

Improving the customer's business should never stop. Gently help the customer understand that.

The Relationship Between the Success Outcome and a Project Outcome

The Success Outcome describes the to-be state you help customers achieve and then improve over time. But the Success Outcome will never be 100% achieved by a customer.

A project outcome is the improvement in the Success Outcome you will make with a business initiative.

There will always be room for improvement in a Success Outcome. Let's look at a couple of examples of Success Outcomes from earlier in the book:

- ERP software—companies buy ERP software to help them run their operations effectively. A Success Outcome for an ERP vendor could be 'Effective Operations'. But customers' operations will never be perfect. They'll always be looking for ways to reduce cost or improve customer service, etc.

- Marketing Automation Vendor—should aim to create Closable Pipeline. But every customer wants to grow. That means the customers' sales targets will go up every year. And so their Closable Pipeline will need to increase every year.

Business initiatives will improve the Success Outcome. For example, an ERP vendor and a customer may undertake a project to reduce operating costs by 3%. That will improve the 'Effective Operations' of the customer.

Or the marketing automation vendor will run a project to help a customer increase Closable Pipeline by 10%. Again, a measurable improvement in the Success Outcome.

References and Advocates

Every project should have a measurable business objective for the customer. But every project should also have an objective for the vendor—to create a reference or advocate.

To win more sales, you need customers who will attest to your outcome-enabling ability. When you first launch your outcome program, you can get away without references. You can talk about the newness of your program. And customers who like what you're doing will act as early adopters. They will undertake an outcome project without checking other customers.

As your outcome program gathers momentum, you will encounter customers who need more reassurance. By this time, you need references.

So, every outcome project should create a reference. And hopefully an advocate.

Section 2 – Products Requiring Implementation

This section deals with more complex products which need implementation services.

Changes for Outcome-based Projects

Vendors of products requiring implementation should consider the following changes.

Measurable Objectives for The Project

B2B companies have not always been great at setting measurable objectives for projects. They focused on getting their products live. Both the vendor and the customer project team were complicit. Neither wanted to have project results placed under the microscope.

So, both celebrated on go-live. The extent of measurement was simple—was the project on time and on budget? The customer may not have gotten great business results. But the vendor fulfilled their obligations.

Things have changed. Now, if the customer doesn't feel successful, it will cost the vendor future revenue. The only way to be certain the customer achieved success is to set objectives and then meet them.

That objective must be measurable. No waffling. Objectives such as 'improved customer satisfaction', 'increased market awareness' or 'increased employee satisfaction' won't work. You need a specific measure. For example:

- Increase average spend per customer by 20% resulting in a profit increase of $20 million.
- Increase sales from marketing-created leads by 15% resulting in increased sales of $100 million.
- Reduce staff turnover by 10% resulting in a $2 million saving in recruitment and training costs.

Clear, specific and measurable—and they should feel achievable. They may present a stretch. But they mustn't feel so difficult that the joint project teams will give up from the start.

Where possible, the objectives should link to one of three core objectives:

- Increased revenue
- Decreased costs
- Reduced business risk

Joint Responsibility

In the past, customers expected vendors to lead product-focused projects. Customers viewed vendors as the product experts. The vendor should know how to make the products work. Customers blamed the vendor for any failures.

With outcome projects, the focus must include all five elements of PSPSPS. The vendor cannot lead some of these elements. In those elements, the customer must take the lead.

Now, the delivery of a business result belongs to both the vendor and customer together. Both must accept a joint responsibility for the success of the project.

Once both project teams accept this approach, they collaborate more effectively.

Project Review Meetings

Every project meeting should have a specific agenda item—is the project on track to achieve a measurable business result? If the project teams feel that's not the case, they need to take corrective action. Even if the project will go live on time and on budget. On time/on budget should no longer be an acceptable measure of project success. It's important, but not enough.

Project Steering Committee Meetings

Both customer and vendor will have executive sponsors. The joint project teams must prepare a single report to present to the executive sponsors.

Most projects run into problems along the way. You cannot allow the separate project teams to blame each other. And they must not present separately to the executive sponsors. One report, one presentation.

It's possible the two teams may disagree. This should be rare. If it occurs, each project team may present their perspective to the joint executive sponsors. But the executive sponsors should take a harsh approach if the project teams could have resolved the disagreement. Strong collaboration between the project teams increases the chance of success. Executive sponsors should ensure collaboration takes place.

Go-live

Everyone should see the go-live as an important milestone. But not as the end of the project. The project should not close until results have been measured and reported to the sponsoring executives.

So, a small celebration on go-live makes sense. It's important. But no one should have any illusion that's the end of the project.

Change Management

The project plan should include a formal change management plan. All participants should have a clear understanding of the to-be state expected from the project. They should all feel clear about why they are making the changes. And how the business will operate.

When the rollout begins, the rest of the company should gain insight into these same topics.

Measurement

Before commencing the project, you and the customer should agree:

- The business objectives for the project
- How and when measurement will take place
- Who will do the measurement
    - you should play a part, and the customer should pay for your involvement

- What baselines you will use for comparing results achieved to past results

Reporting to Sponsors

The joint project teams should prepare and present a joint report to the sponsoring executives. Regardless of whether they achieved the objective.

Your Success Consultant may lead the presentation. But the sponsoring executives must understand it's a joint presentation.

If the project results did not meet the objective, the teams should present a joint analysis. Neither should attempt to allocate blame. It's highly likely both customer and vendor could have done better. Both sides should present lessons learned and talk about how they could improve their own performance.

Where possible, the joint teams should offer options on what could happen next. Remediation will be one option, but others may also exist.

If the project met or exceeded the project objectives, there's an opportunity to create heroes and heroines in the customer. The vendor should give credit to the Decider and the customer team.

You may find some initial reticence from your professional services teams. We cover this in Appendix 2.

Section 3 – Products Not Requiring Implementation

Simpler products may not need implementation services. They'll drive less business change than their more complex cousins. Their adoption should prove rapid and users must like the engagement immediately. Think about apps you download on your phone.

But the simplicity doesn't mean the other elements required to achieve an outcome remain unaffected. An outcome still requires PSPSPS.

The difference lies in how you execute the PSPSPS changes. With complex products, the project should address all PSPSPS aspects. The product will take care of some of these elements. You'll handle the rest as part of the project.

With simpler products, the product itself must deliver most of the PSPSPS aspects. You may include:

- A guide to the best process for achieving the business outcome your product supports
- The reasons a company would adopt the product
- Built-in training
- Built-in coaching—the system identifies where an individual user may be struggling and offers coaching tips
- Automated measurement and reporting of the business outcome

Product-Enabled Outcomes

As a general rule, the simpler the product, the more of the outcome the product should enable.

However, more complex products should also strive to deliver more of the outcome through the product. The simpler and easier you can make achievement of the Success Outcome, the more likely that achievement will occur.

And that will drive revenue growth.

Summary

With subscription pricing, the initial sale is not as significant as it once was. Now, customers commit to the smallest subscription possible. And they will only grow the subscription if they feel successful.

Vendors should focus on enabling a business result which higher management of the customer will regard as a success. With success, the customer will grow their commitment to you.

For products requiring implementation services, your project methodology should evolve. You and the customer will agree a measurable business objective for the project. Throughout the project, both your team and the customer's team will work together to achieve this objective. You'll treat go-live as an important milestone, but not as the end of the project. You'll measure results achieved after a suitable 'settling-in' period. And report the results back to the Decider(s).

For simpler products, the product itself should deliver more of the outcome. But even vendors of more complex products should aim to deliver more and more of PSPSPS through the product.

You should aim for Product-Enabled Outcomes.

PART 4 – Advocates Created

Every business needs references. You can't sell without references, except in the start-up phase of the company.

Many (perhaps most?) vendors face a tough reality. The bulk of customers will say something positive about a vendor, such as 'They're OK.' But most won't strongly endorse the vendor. Often vendors have only a handful of good references. And can sometimes overwork that handful.

All companies wish they had more advocates. There's no better marketing than a customer offering unsolicited praises.

In Part 4, we'll cover how to set up an outcome-driven advocate program.

Section 1 – References and Advocates

What are references and advocates?

- Reference—a customer where at least one person will take a call from a prospect or other customer to discuss their experience with you as a vendor. They will say positive things, but only when asked.
- Advocate—a person who proactively tells other people how great you are to work with. That includes telling people in their company as well as friends and associates outside of their company.

Hailed

The fifth decision, Decide to Advocate, requires customer champions to feel victorious at the end of the project. And then for their company to hail them as heroes and heroines.

You should help ensure that both happen.

Section 2 – A Virtuous Circle

You can use your outcome program to develop a formal advocate program. Don't leave the creation of references and advocates to chance. To do this, you're going to create a virtuous circle.

Enable and Measure the Success Outcome

Creating advocates begins with helping the customer achieve, or improve, the Success Outcome. You helped them create an ongoing business result which higher management of the customer will regard as a success.

Use this success as the centrepiece of your advocate program.

Create Heroes and Heroines

You should ensure key customer staff get credit for the great result. This includes your Decider and the key members of the team around them.

Take positive steps to ensure others in the customer, particularly higher management, take notice. For example:

- Have an award ceremony with higher management present. Give the Decider and team a certificate or plaque acknowledging their success.
- Send a letter from someone senior in your organisation to their higher management congratulating the key people.
- Record a video capturing their success. Distribute it internally in their company. You may also consider publishing it externally, for example, on social media.

Your Decider and team will enjoy the praise higher management will give them. They may simply receive recognition. But perhaps it means a bonus. Or even a promotion.

Higher management will hail them as heroes and heroines.

Heroes and Heroines Become Advocates

You helped the Decider and team receive recognition in their company. They know they could not have achieved this without you. And they are grateful for you assigning credit to them.

It's likely they will reciprocate. They will proactively sing your praises to other people, both within their company and to friends and associates. They will become advocates. Or they will at least act as references.

Advocates Ensure Retention

When it comes time for renewal, your advocates will ensure renewal goes ahead. After all, you've linked their success to your success. They don't want to see you dropped or replaced. They will lobby hard to keep you if it's ever questioned.

Advocates Drive Expansion Revenue

Your Decider sponsored a business initiative. It created a business result higher management of the customer regards as a success. Through your efforts, you ensured the Decider and the team received credit. They became heroes and heroines.

A year later, perhaps earlier, you can jointly identify the next improvement in the Success Outcome. That will lead to another business initiative. Your advocates know if another business initiative proves successful, you will again ensure they get credit.

So, your advocates will sponsor the next round of improvements of the Success Outcome. They're keen to see another project that will lead to reinforcing their status as heroes and heroines.

Your virtuous circle looks like this:

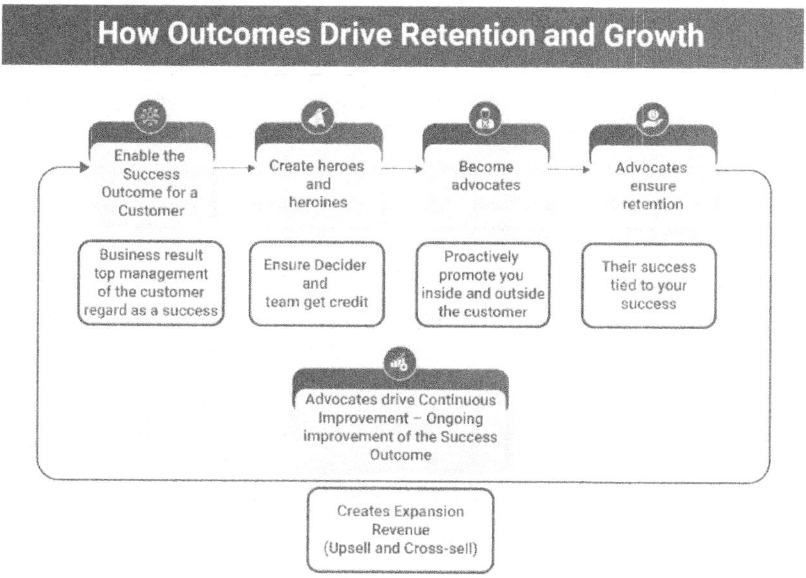

Figure 66: Virtuous circle

PART 5 – Cycle Looped

Section 1 – Repeating Cycle of Engagement

Your IDEEAL Lifecycle creates a cycle you want to repeat at least every year, sometimes more often.

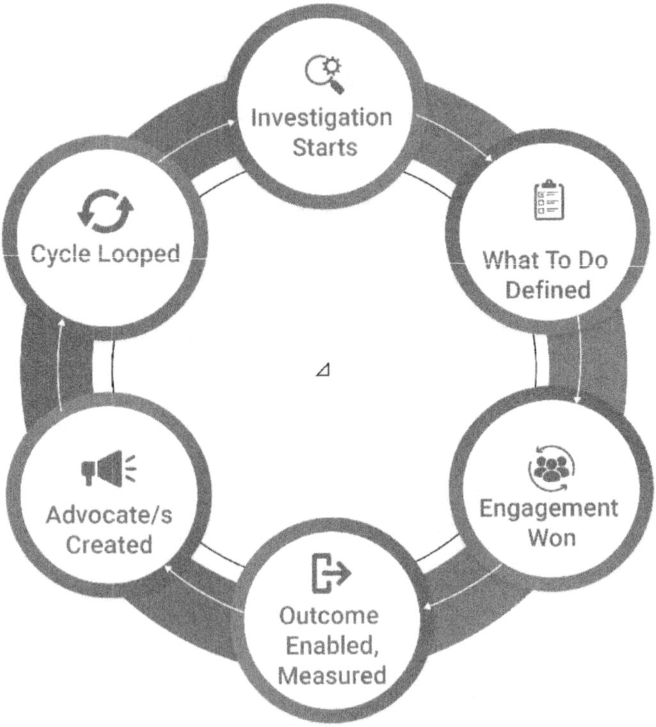

Figure 67: IDEEAL Lifecycle

As discussed, for your Deciders to Decide to Loop, you need them to anticipate again becoming heroes and heroines.

You should formalise how you will work with them to make them heroes and heroines again. You will establish a repeating cycle of engagement. You can cover topics such as:

- When and how will measurement of the Success Outcome and Contributing Outcomes take place?
- Who will do the measurement?
- Who will create the reports?
- Who will receive the reports?
- How will the next joint review process take place? For high-touch customers this will mean the next Executive Business Review (EBR).
- Who will take part in the next review process, including the EBR?
- How will preparation for the review process take place?

Catalysing the Next Business Initiative

To restart the IDEEAL Lifecycle, you kick off another round of your BuILD process. For another business initiative to start, you need a Decider to Decide to Investigate. You want to catalyse that decision to act.

That's how you drive expansion revenue.

Primary Lead Generation Method

Your BuILD process should become your primary method of generating expansion opportunities. Engage with Deciders at least once a year. Discuss ways to improve the Success Outcome.

Through this approach, you will uncover most of the opportunities for business initiatives. From time to time, unexpected opportunities will arise. But for most customers, you will have visibility of all possible business initiatives.

Medium-term View

The BuILD process will create some business initiatives that begin right away. For others, the customer will decide to defer a decision. These other business initiatives may have a good business case, but the timing isn't right.

The deferred business initiatives create a medium-term pipeline. Over time and across all customers, you will build these medium-term opportunities. This helps with medium-term budgeting and capacity planning.

PART 6 – Outcome Techniques

Earlier in the book, we talked about what your customers must get right to achieve a business outcome—PSPSPS. It's in your interest that customers get all five elements right.

To assist, we introduce a new role—Success Consultant. This person understands all five PSPSPS elements and will help customers get them right.

We will also provide more detail on how to work with the customer on a plan for a business initiative. And how to be paid for this work. We call this the 10% Sale.

A challenge in many sales situations lies in the number of prospect or customer people involved in the buying cycle. It's difficult to get them aligned. In Part 6, we'll offer a method for creating the needed alignment.

And we will also offer a steppingstone for moving customers and prospects into your outcome program.

Section 1 – Success Consulting

Success Consultants understand what's needed to enable the Success Outcome. They know about all five elements of PSPSPS. They help ensure the customer gets all five elements right.

Success Consultants will engage with customers at key points in the customer journey. They may assist Deciders in Deciding to Investigate. They'll play a key role in the customers' Decide What to Do phase. They will engage during the Decide to Engage You phase. And they will play a quality-control role during Outcome Enabled, Measured. As a result, they provide continuity of engagement across the lifecycle.

This will help you fulfil your revenue potential from each customer.

How Things Have Changed

In the past, vendors didn't need to focus on all aspects of PSPSPS. They received most of the money from a customer upfront. As long as the product worked according to specifications, the vendor considered their job done. And they were quick to avoid accountability for anything beyond the immediate scope of their products and services.

Some customers knew how to get all elements of PSPSPS right. And their projects succeeded.

But many customers lacked experience in this type of project. These projects had a high risk of partial or complete failure. But vendors did not accept accountability. These failed projects left the customers feeling bitter, but for the most part, helpless.

As we discussed, subscription pricing has shifted power back to customers. Vendors now have a vested interest in ensuring the customers feel successful. And that's why vendors need Success Consultants.

Characteristics of Success Consultants

For a vendor, Success Consultants will be the most senior consultants in the organisation. These consultants will have broad industry experience.

And the ability to link customer problems to your Outcome Structure. They will guide the customer in development of a plan to achieve the Success Outcome. And then help the customer improve the Success Outcome over time.

Success Consultants should impress customer executives. Their personal charisma will help establish an initial relationship. Their knowledge in enabling your Success Outcome will make them advisers to the customer executives. And that's what will keep you engaged at this level.

Success Consultants will have strong communication skills, both written and verbal. They'll have strong analytical skills. And the ability to apply their knowledge and experience to create great outcomes for customers.

They don't need to be experts in your products. They need to know enough to ensure your products can deliver any recommendations they make. But that's enough. Sometimes, it's better they are not product experts. That way, the customer isn't tempted to ask them detailed questions about product functionality. You will have product experts who can provide that kind of information.

Success Consultants will also be strong teachers and mentors for other staff. They'll lift the knowledge and skills of those around them. And that will help embed an outcome culture in your business.

Why Customers Like Success Consulting from Vendors

More Cost Effective than Internal Customer Resources

Success Consultants have in-depth knowledge and experience. Their knowledge and experience increase success at key points in the customer's journey. But that knowledge and experience aren't needed all the time.

Good Success Consultants are not cheap to hire. If a customer hired a Success Consultant internally, they could not use the Success Consultant's skills all the time. The customer will have invested in an expensive resource they aren't fully utilising. That's not good use of the customer's budget. And the under-utilisation would make it difficult to retain good Success Consultants. They'd get bored and leave.

It's more cost effective for a customer to outsource this expensive resource. Vendors can apply the in-depth knowledge and experience of the Success Consultants across many customers. Each customer need pay only for the time they need that in-depth knowledge and experience.

Better than Independent Consultants

Not Reliant on Billing Opportunities

Independent consultants earn their income from consulting activities. They try to maximise the amount of consulting a customer needs. They want to make every engagement as large as possible.

Product vendors do not rely on consulting revenue as their primary source of revenue. Success Consultants play a strong role in delivering product revenue. So, vendors want their Success Consultants helping as many customers as possible.

For an individual customer, Success Consultants will engage for the smallest amount of time possible. And then move on to the next customer. This will keep the cost of Success Consulting to a minimum for each customer.

'One Throat to Choke'

When something goes wrong with a project, everyone hopes to avoid accountability. If the project has both an independent consultant and a vendor involved, they tend to blame each other. And the customer flounders in the middle, wondering where the truth lies.

If a customer uses a vendor's Success Consultant, there is only 'one throat to choke'. If something does go wrong, the vendor must work with the customer to rectify the situation.

Practical Recommendations

Independent consultants can sometimes make recommendations that a vendor cannot support. This puts the customer in a difficult position. They can ignore the independent consultant's advice and follow standard practice in the vendor's products. Or they can undertake expensive customisation. Either way, there's confusion and added expense.

Success Consultants don't need to be product experts. But they need to know enough about the product to ensure the vendor can support any recommendations they make. This means Success Consultants will always make practical recommendations.

Why Offering Success Consulting is Good for You (the Vendor)

You will use your outcome program to differentiate from your competitors. Your competitors focus on product outcomes or solutions. You focus on the Success Outcome you enable. You emphasise the importance of PSPSPS.

Success Consultants send a message to your customers—you're focused on meaningful business outcomes for them. Your outcome approach is not a clever sales ploy. It has substance.

This will help you win more deals.

Your company will find Success Consultants invaluable inside the company. Their in-depth knowledge and experience will help raise the skills of the staff around them. For example, they will:

- Advise marketing on campaigns.
- Play an integral role in the sales process.
- Play a high-level role in projects, ensuring your team delivers the business outcome discussed during the sales cycle.
- Advise product management and development.

Success Consultants will also generate direct revenue. Much of the time they engage with customers should involve paid engagements. And their day-rate should be high. Think double your standard product-consulting rates. They will be expensive to hire. But with the right day-rate, they can be self-funding with as little as 80 billable days a year.

Most importantly, they will increase the likelihood of success for your customers. And that increases the likelihood of making sales.

Tasks that Success Consultants Lead

In each step of your outcome lifecycle, Success Consultants can play a role:

Investigation Starts

For your face-to-face customers, the Success Consultant will play a lead role in the Executive Business Review (EBR). They'll lead the conversation on industry trends. They will both lead and provide analysis of the results the customer has achieved. They'll play a strong role in influencing Deciders to Decide to Investigate.

For self-help customers, they will play a role in designing the engagement model for this segment of customers.

What to Do Defined

For your face-to-face customers, the Success Consultant will lead the analysis of business initiatives.

For your self-help customers, Success Consultants will help design the tools you offer. These tools will allow self-help customers to develop a plan for a business initiative.

Engagement Won

Success Consultants can play an active role in ensuring the customer's selection criteria focus on PSPSPS. They can remind the customer why selection should not focus on product features and functions.

Outcome Enabled, Measured

Success Consultants will provide a bridge between your sales teams and professional services teams. They played a lead role in developing the plan for a business initiative. They have an in-depth understanding of what the professional services team needs to deliver.

During a project, they will focus on quality assurance. They will ensure the project stays on track to deliver the business objective.

Product Development

Success Consultants provide an excellent resource when considering your product roadmap. They should provide input on the best way to improve a vendor's ability to enable the Success Outcome.

Summary

As part of your outcome program, Success Consulting will prove one of the best investments you make. It may stretch your budget, but you should promote or hire the best people you can for this role. It will make a huge difference to your success.

Redseal helps organisations achieve cyber resiliency. Ramesh Kaza, Vice President of Customer Success said: We have a team of consultants who help customers with everything needed for customers to become cyber resilient.

Cyber resiliency requires three Contributing Outcomes. First, our customers need to achieve visibility. They need to know what's on their network and what it's being used for.

Next, customers need to know their exposure. For example, they need to know the potential attacks and the likely impact on the business. Our consultants discuss cyber standards and help the customers understand their compliance with those standards. Compliance is the second Contributing Outcome customers need.

The third Contributing Outcome involves risk management. Cyber resiliency isn't black or white. Some organizations need stronger protection and resiliency than others.

Our consultants guide customers to assess the level of resiliency they need. They help the customers develop a plan for attaining that resiliency. And a plan for how to maintain that resiliency over time.

Throughout this process, our consultants remain focused on the three Contributing Outcomes needed for resiliency—visibility, compliance and risk management.

Their engagement focuses on everything the customer needs to achieve resiliency. It's all about the customer's business outcomes, not RedSeal's products.

Section 2 – The 10% Sale

The 10% Sale results in a deep study on a business initiative. It creates a detailed plan for that business initiative. The detailed plan has everything needed for the customer to feel confident about the business initiative. And later, to obtain approval for the business initiative.

It's a technique for obtaining a significant commitment from a customer. It's the perfect culmination of your BuILD process. But it also plays a strong role in ad-hoc opportunities within existing customers.

In most circumstances, customers should pay for your involvement in creating a detailed plan. You're not doing a 'pre-sales' activity designed to position your product. It's a deep study covering everything needed for the business initiative.

The more complex your products and the outcomes you enable, the more useful a 10% Sale will prove. For simpler products and outcomes, you may create automated versions the customer can use through a self-help model.

For the rest of this section, we will treat the 10% Sale as a face-to-face engagement.

Key Features of the 10% Sale

The 10% Sale creates a detailed plan. The detailed plan contains the following:

- Description of the to-be state to attain from the project. Everyone involved needs a clear idea of how the business will operate when the business initiative succeeds. As Stephen R Covey advised us in *The Seven Habits of Highly Effective People* (Covey, 1988), we should start with the end in mind.
- The detailed plan will cover all aspects needed to achieve the to-be state—PSPSPS.
- The plan should include a description of what will need to change to achieve the new to-be state.

- The plan should remain product agnostic. The customer could use it with any vendor. Your ability to charge for this consulting work depends, in part, on you remaining product agnostic. Also, you're still in the customer's Decide What to Do phase. Discussion of your product isn't phase-appropriate. You'll lose the customer's faith if you talk about your products.
- The plan will include a full financial analysis. The format will follow whatever the customer needs to get approval for the business initiative. It's in your interest to ensure the plan contains everything needed for approval. And it needs to be clear, concise and easily understood.

Positioning the 10% Sale

Imagine you have conducted an Executive Business Review with a customer. The customer executives identified three possible improvement areas. You and key customer staff then conducted a high-level analysis of each possible improvement area which showed:

- The first improvement area did not have a strong business case. You and the key customer staff jointly recommend against proceeding.
- The second improvement area had a mediocre business case. The time and cost required to improve that area and the likely benefits look balanced. It's unlikely the customer would proceed.
- The third improvement area showed a strong business case. You recommend going ahead.

You jointly present these findings to the customer executives. They agree they should not proceed with the first improvement area.

On the second improvement area, they accept it has a mediocre business case. But the board wants something done. The executives decide to keep this open, but not proceed immediately.

They accept the strong business case for the third improvement area. They can see they need to act. They ask what the next step should be.

You say: 'You will need a detailed plan for how to achieve the improvements in the business. We recommend that we, along with your key staff, conduct a study.

We will jointly write a detailed report that includes everything you would need to achieve success.'

The senior executive says: 'Why should we do a study? Why don't you just give us a proposal?'

You respond: 'The detailed report will contain everything you need for approval of this business initiative. It will give you complete confidence before making any further commitments to the business initiative. And the report will become your submission if you decide to proceed.

'Would you like us to submit a proposal and costing for the study?'

A look of surprise crosses the executives' faces. Up to now, you have not charged for the work you have done.

The most senior executive asks: 'Are you saying we need to pay for this study? Why should we pay?'

You respond: 'The detailed report we will jointly create is not a sales proposal. It's a deep study on how to improve this area of your business. We will cover all elements required to achieve the better business results. We'll cover processes, systems, people skills, partner involvement and supporting elements such as policies, measurement and reporting.

'The systems component will be product agnostic. We will not reference our products or services. We will describe any system changes you need in a generic way. It's a report you could use with any vendor, not just with us.

'In fact, it's possible you may not need any investment in systems. If that's the conclusion our joint teams reach, that's what we will say in the report.

'For these reasons, this is a paid consulting activity. It's not a sales activity.

'But there's good news. If you go ahead with the business initiative, doing the study first will have little impact on the total cost. Let me explain why.'

At this point, you bring up a slide with this diagram:

Figure 68: Project steps without the 10% Sale

You continue: 'In the past, the typical steps in a project included:

- Design—design the way the business should run at the end of the project.
- Plan—develop a project plan that sets out the steps required, the timing, interdependencies, etc.
- Kick off—project teams and executive sponsors from both customer and vendor work together. They confirm the scope, timing, processes, review methods, governance, etc. for the project.
- Execution—the detailed work of the project
- Live—go-live of the product
- Close—we conclude final paperwork and administrative matters and close the project.

The star indicates when you approve the full project. In the past, you would provide final approval before any of these activities start.'

You now bring up a second slide:

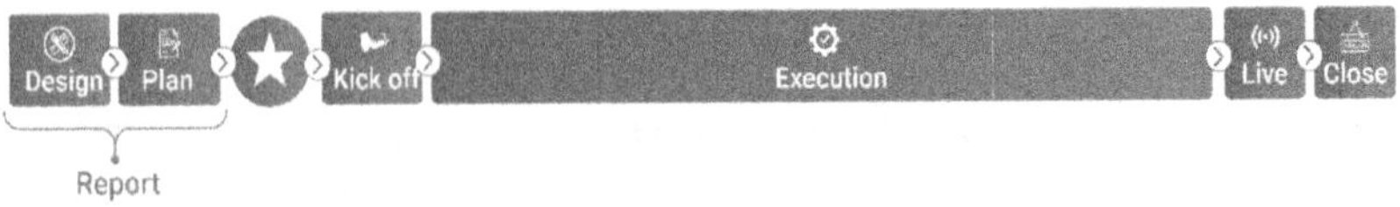

Figure 69: Project steps with the 10% Sale

And you say: 'When you do a study first, the project will look like this new diagram.

'The study brings forward activities that would normally occur at the beginning of a project anyway—the Design and Plan steps. But now we will do these activities before you commit to the full project.

'You can see the star, indicating your commitment point, has moved.

'This reduces risk and increases alignment around what the project must deliver.

'Here's an example of reports we've produced for other customers.'

You hand over two example reports, redacted to hide any customer details.

The executives pass the reports around and flick through them. Impressed, the senior executive says: 'OK, give us a proposal for this study. It sounds like a sensible approach.'

When to Use the 10% Sale

BuILD Process

You should use the 10% Sale in the Plan step of your BuILD process.

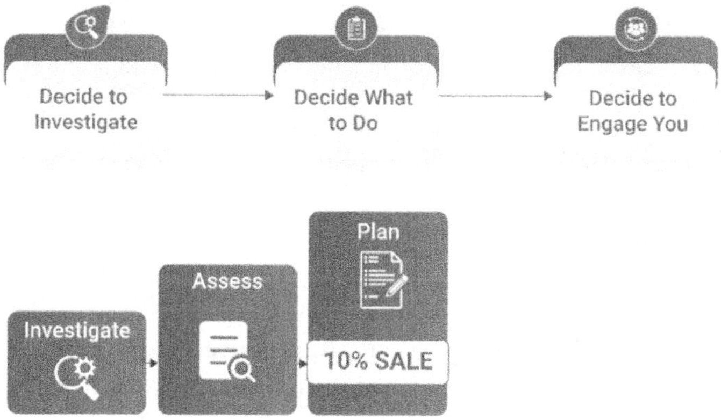

Figure 70: Where the 10% Sale fits in the BuILD process

The Decider has decided to investigate. You've created a high-level plan for success. You've then obtained approval to proceed with the detailed study. The 10% Sale is that detailed study.

Ad-hoc Opportunities

With existing customers, ad-hoc opportunities will arise from time to time. Perhaps a customer has a new problem which requires immediate attention. Or there's a change in their company policy. Or an unexpected external influence. A Decider will Decide to Investigate.

The Decider will then need to Decide What to Do. They will need a plan for a business initiative. And they will need to ensure a strong business case exists. The 10% Sale provides the perfect solution.

In other words, whenever an ad-hoc opportunity arrives, propose a deep study through a 10% Sale. You'll find customers will reach a point where they start requesting these deep studies.

New Business (New Logo)

You can use the 10% Sale to disrupt and then take charge of a prospect's buying cycle.

In buying cycles today, prospects engage vendors late in the buying process. They're expecting to go through an evaluation of your products and services.

You'll attempt to disrupt this process by bringing them into your BuILD process. And you'll use a concept called a Business Malady to effect this disruption.

We'll discuss Business Maladies in Section 4 of this part.

Price for the 10% Sale

As a rough rule of thumb, the price for this detailed study should be about 10% of the total project cost. You can produce a rough estimate of what you will charge the customer for this business initiative. Charging about 10% of this amount for the study will feel comfortable for most customers.

So, if the full project cost would be $100k, you'd charge $10k for the study.

Benefits for the Customer of the 10% Sale

Lowers Risk

With a detailed report, the customer has much more insight into the project before making a final commitment. It's now far less likely they will get unpleasant surprises.

When it's time to seek final approval for the business initiative, the customer executives can do so with far higher confidence.

Increased Likelihood of Success

The study will work through potential issues and ensure alignment of all parties. The clear description of the expected to-be state will give everyone a common focus.

An Easier Next Step

The 10% Sale normally follows the high-level Assess step in BuILD. It's a smaller step compared to a commitment for the full business initiative. As a result, it's an easier commitment for customer executives to make.

Benefits for a Vendor

Qualification

The 10% Sale provides an excellent qualification tool. If the customer won't commit 10% of the full project cost to get a detailed report and plan, they probably aren't committed to the project.

In the past, you would do lots of demonstrations to convince the customer to move ahead. Making the 10% Sale qualifies the customer. And avoids lots of wasted energy and time.

Defeats 'Do-nothing' Decisions

'Do-nothing' decisions often prove the biggest competitor vendors have. Customers go through what appears to be an evaluation with a vendor. At the end, they decide they will not move ahead with a business initiative.

If they commit to the Outcome Study, they have made a financial commitment. It's not as easy for them to decide to do nothing.

High Conversion Rate

Unless the customer feels it's likely the business initiative will proceed, they will not spend the 10% for the study.

As a result, you can expect conversion rates above 90%. In other words, almost every customer who commits to a study will also commit to a full business initiative.

Increased Likelihood of Success

Customers don't always have clear expectations for a project. If that isn't rectified, the project runs the risk of meandering. And that increases the likelihood the customer will not feel it's a success.

The study will ensure clear expectations for all parties. It will describe the to-be state, the roles everyone will play, timelines, costs and so on.

With clarity and alignment around all project elements, the likelihood of success increases.

Shorter Sales Cycles

For a business initiative to proceed, the customer must have confidence in a plan.

The 10% Sale has everything needed for the customer to achieve confidence to move forward. This minimises delays due to indecision.

The study also minimises the need for lengthy demonstrations. The customer learns how the business will run from the report, not from product demonstrations. If the customer wants demonstrations, they tend to be proof-of-concept rather than in-depth studies.

Paid 'Pre-sales'

The 10% Sale can reduce your cost of sale. You would normally undertake lengthy demonstrations and the writing of sales proposals. You do all this work for free. It's time-consuming and expensive.

The 10% Sale delivers the bulk of work required to win a deal. And you're paid to carry out the work. That can reduce your cost of sale.

Productising the 10% Sale

The 10% Sale does not need to be bespoke for every customer. Possible improvement areas for customers will often prove similar. If you find improvement areas that surface regularly, you can develop a standard proposal for a study in that area.

You can also use this standard proposal for proactive selling of outcome studies.

Summary

The 10% Sale completes your BuILD process. It completes the building of confidence needed for a customer to proceed.

It's also a useful tool when ad-hoc opportunities arise. And can be used for New Business (New Logo) opportunities.

The 10% Sale creates a report with everything the customer needs to get approval for a business initiative. It's an easy step before committing to a full project for a customer. It reduces the risk and increases the likelihood of success for them.

For the vendor, it's a great qualification tool. If customers commit to an Outcome Study, most will proceed with the full business initiative. It's a more effective way of moving customers along their buying cycle, eventually leading to a sale.

Section 3 – Dealing with Many Deciders in a Deal

If your products and services impact many departments, you may engage more than one Decider when working on a deal. For example, each department head may have a strong say in whether a business initiative goes forward.

In *The Challenger Customer* (Adamson et al., 2015), the authors describe the difficulties of dealing with many people. A salesperson will attempt to understand the needs of each person. But each person has a unique set of requirements. The common ground diminishes as you engage each new person. The requirements everyone can agree on become smaller and smaller. And that drives down how much the customer will pay to achieve that ever-diminishing set of requirements.

Your Outcome Structure provides a great tool for dealing with many people in a deal.

You don't drill down to the unique requirements of each department and try to find common ground. You establish the common ground at a level above each department. Your Success Outcome provides the key.

An Example

Consider an ERP software vendor selling to a manufacturing company.

Earlier, we suggested a possible Success Outcome for an ERP vendor—Effective Operations. Manufacturing companies buy ERP software to get their operations running as effectively as they can. The ERP vendor helps them achieve Effective Operations. That's the to-be state they aim for.

An ERP system affects every department of a manufacturing company. So, there'll be lots of Deciders involved.

Let's imagine an ERP salesperson, Jill, has a meeting with the head of production, Ravi.

Jill and Ravi exchange pleasantries, then get down to business. 'Ravi, what changes do you feel the business needs?' says Jill.

'We're a manufacturing company, so it's clear that my production department is the most important,' says Ravi. 'The company should focus on my needs.'

'OK, can you describe those needs?' says Jill.

'There's a long list, but we have two key needs. First, we're getting a backlog of inventory at a few key workstations. That's slowing down the production process. We need a scheduling tool to better plan work at the key workstations.

'Second, we're getting too many product specification changes. What we're making keeps changing. We need stability in product specifications.'

Jill could already see a conflict because she had already interviewed the head of product engineering. She knew the changes in product specifications aimed to bring down product costs. But the focus on cost was adversely impacting the production department.

Rather than tackle the conflict head-on, Jill takes a different approach.

'Those two issues sound like they'd have a big impact on production. And also on whether your company achieves Effective Operations,' says Jill.

'You're right about the impact on production. But what do you mean by Effective Operations?' says Ravi.

'Manufacturing companies aim to provide what customers want, when customers want it, and at the right quality and cost. That requires the overall operations of the company to run effectively.

'We find that every department in a manufacturer focuses on its requirements. And it's hard to see the perspectives of other departments. So, we try to find a focus that will produce the best overall result for your company. We call the to-be state we aim for Effective Operations,' says Jill.

'OK, makes sense, but how does that affect me?' says Ravi.

'We have a list of requirements from every department. And there are some conflicts and differences of opinion. We're going to recommend to your executive committee, which includes you, that we jointly carry out a study. The study will identify the best way to improve the effectiveness of your operations.

'You and your fellow executive committee members will probably have to set priorities. We will suggest one question to help you choose between options. We'll ask: "What will make the biggest difference in achieving Effective Operations?"

'That single question will give everyone a common focus. It will be what your CEO will want. And that will increase the chance of reaching an agreement on the best way to make a positive difference for your company.

'Does that sound like a sensible approach?' says Jill.

'I guess so. But are you saying I might not get what I need?' says Ravi.

'You and the rest of the executive committee will make that call. You might agree to a compromise on one of your needs if you can see that the overall operations will be more effective. Again, that's probably the attitude your CEO would hope for,' says Jill.

'OK, I can see how that could work,' says Ravi.

Jill has skilfully lifted Ravi's sights. Instead of a focus only on the needs of the Production Department, Ravi now has a frame of reference for considering the good of the company.

The ERP vendor's Success Outcome has allowed a unified perspective for the customer. They can now decide the scope and objectives of a business initiative with a united perspective.

The good news for Ravi is that Jill will also suggest a continuous improvement approach. Every year, Jill's company and the customer will jointly review the operations of the business. They'll identify the next improvements they can make in Effective Operations. That will lead to another business initiative. If Ravi doesn't get everything he wants this year, perhaps it will rise to the top of the priorities next year.

And even if it doesn't, he and his fellow executives will understand why.

Section 4 – Moving Prospects and Customers to an Outcome Focus

You want your prospects and customers to focus on your Success Outcome. But prospects and customers will often focus on the problems or challenges they face. You need to help them make the transition from a problem focus to an outcome focus.

The interesting side benefit to an outcome focus is that problems disappear. Consider a medical analogy. A patient has a series of physical symptoms. But they are also suffering from depression because of their medical condition. The doctor diagnoses their medical condition, and treatment restores their health. Not only do the physical symptoms disappear, but the depression also melts away.

Achieving an outcome causes problems to disappear. You can use this idea to help customers transition from a problem focus to an outcome focus.

Clarity Around the Problems You Solve

We discussed the concept of an Outcome Hierarchy. Each role in a company delivers a single core outcome for that company. People in each role will face a series of problems or challenges in achieving that outcome. In companies in the same industry, people in the same role face similar problems. The problems don't change much from company to company.

It's possible to create a list of problems and challenges faced by people in a specific role. On an Outcome Hierarchy, it might look like this:

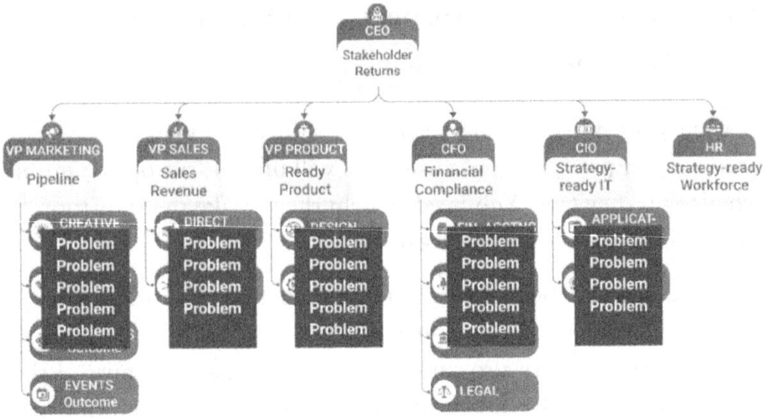

Figure 71: Each role in a customer faces a set of problems

Your Problem Set

You have chosen where you fit in the Outcome Hierarchy. You know who your Decider is, and thus where your Success Outcome aligns. You can now list all the problems a Decider would face in achieving the Success Outcome.

But you won't solve every problem a Decider faces. For example, your Decider might be the Chief Financial Officer. They can face a conflict between strict adherence to accounting regulations and the company's desire to book as much revenue as possible. Those two things aren't always compatible. You probably can't help the CFO resolve this internal conflict.

From the full list of problems, select those where you can have a real influence. They will have a link to the Success Outcome. You now have the problem set you help prospects and customers address.

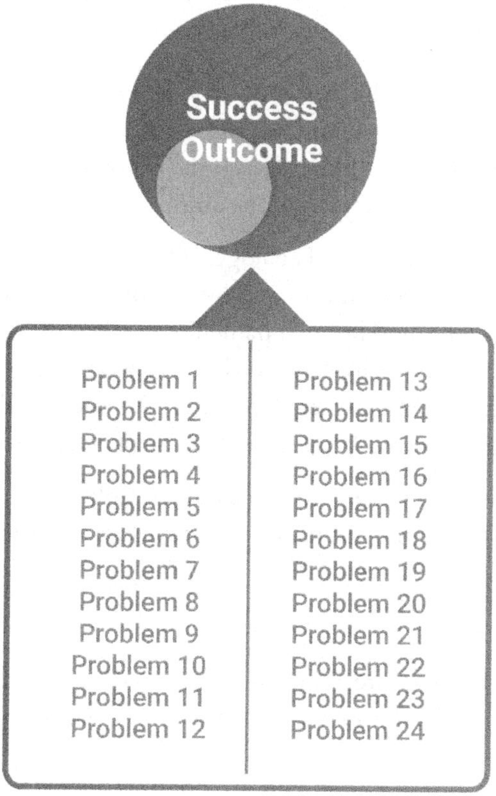

Figure 72: Your problem set – the problems you can solve for customers

When you help a new customer achieve the Success Outcome, you'll solve these problems.

To be fair, you won't solve them all at once. The Success Outcome will never be 100% achieved by any customer. Customers will need to keep working to improve the Success Outcome. And that need for continuous improvement will underpin your expansion revenue program.

A Steppingstone from Customer Problems to Customer Outcomes – Business Maladies

It's possible to simplify the link between a set of problems and the Success Outcome that will solve them. Give the set of problems a label. The label will make establishing the link between the problems and the outcome easier. And it will make conversations on this topic simpler.

Let's continue to use a medical analogy. If we feel unwell, we go to a doctor. We describe the symptoms. The doctor will run some tests and will give us a diagnosis. They will name the medical condition or malady we're suffering from. It's likely we will know something about this malady. The label creates a mental picture. The doctor will explain how to treat the malady, and you'll agree on a course of action.

We do the same thing with a set of customer problems. Think of the problems as symptoms of a Business Malady. Just as doctors name a medical malady, we name the Business Malady.

ERP Software

Consider an ERP software company selling to the manufacturing industry. The types of problems a manufacturing company might face include:

- Inaccurate forecasts
- Purchasing the wrong raw materials
- Production bottlenecks
- Finished goods stockouts
- Delivering the wrong items to customers
- Incorrect invoices

The manufacturer can overcome the problems, but each needs corrective action. The problems drag down productivity and quality, and increase costs.

The ERP vendor wants a single label to describe this set of problems. They might choose something like 'Operational Glitches'. Each of the problems is a glitch in the manufacturing operations, something that's not quite working as it should.

So, the ERP company helps manufacturers fix 'Operational Glitches'. The word 'glitches' implies manufacturers have problems, but they are not insurmountable. And every business has glitches, so a manufacturer isn't likely to take offence.

The Success Outcome for this ERP vendor is 'Effective Operations'. That's the to-be state they help their customers achieve. When a manufacturer achieves effective operations, their operational glitches will disappear.

When the ERP vendor sells to a New Business prospect, they initially talk about 'Operational Glitches'. They discuss the major glitches with the prospect. They may jointly assess the impact of those glitches with the prospect.

The ERP vendor then paints the picture of what Effective Operations would look like. They help the customer see what the future to-be state could look like. And help the customer understand how achieving the to-be state will cause the operational glitches to disappear.

The ERP vendor has:

- Stepped the prospect through a discussion of the prospect's problems
- Introduced a label embracing all the problems—the Business Malady of 'Operational Glitches'
- Assessed the impact of the major operational glitches
- Introduced the Success Outcome of 'Effective Operations'—the to-be state the ERP vendor can help the prospect achieve
- Helped the prospect see how achieving 'Effective Operations' will cause their operational glitches to disappear

They have now moved the prospect from a focus on current problems to a focus on a future to-be state using the steppingstone of the Business Malady.

Now they are ready to help the prospect define a business initiative to achieve Effective Operations.

The Right Time to Use the Business Malady Approach

Experimentation with the Business Malady approach will help you find what works for you.

It is most helpful when dealing with people unfamiliar with your outcome approach. It's a way of transitioning people more used to focusing on current problems. This will include New Business (New Logo) opportunities. And new people in existing customers.

If you're engaging senior executives, some may find the link directly to an outcome more helpful. They won't need the steppingstone of the Business Malady.

No Criticism

In using the Business Malady approach, you don't want a customer or prospect to feel you're critical of their current performance. They may feel you're implying they aren't doing a good job.

It will help to emphasise two aspects. First, you'll explain that lots of companies experience this Business Malady. In discussion, almost all agree they want to do better. Second, you should focus on identifying opportunities for improvement, not criticising current performance. Keep it positive.

PART 7 – Segmenting Your Existing Customers

There is an approach to segmentation that balances the economic return from a customer with the way they want to engage with you. You will learn about it in Part 7, along with suggestions on how to engage with each of the four resulting segments.

Section 1 – Segmentation

You will find no shortage of criteria you could use for segmenting your customer base. You could use customer size, revenue to you, industry, location and many more. With your segmentation established, you can establish two main things. First, you can allocate your resources by segment. For example, you can decide how many account managers to allocate to each segment.

Second, you can decide on an engagement model for each segment.

Two Criteria

We will concentrate on two criteria for establishing segmentation under your outcome program.

The Return to You From Each Customer

The return to you from a customer has three elements:

- Current economic return—most vendors would see revenue as the primary focus. But for businesses wishing to increase profit, gross margin becomes important. It's possible, for example, that customer acquisition cost (CAC) varies between customers. Some customers may require extensive demonstrations, proof-of-concept, reference visits and more before any purchase. These customers have a high CAC, which reduces gross margin. You may de-emphasise customers with a high CAC to improve overall profitability.

- Growth potential—some customers may have a low economic return at present. But they have a high potential for growth. You may choose to apply more resources to these customers to ensure that growth.
- Non-economic returns—every company needs references and would love to have more advocates. You may choose to provide stronger support for those accounts which assist you with reference calls or visits, or who act as advocates. It's a sensible investment. Other customers may have a high brand recognition which will assist in making sales to other companies. Again, you may over-invest in these highly recognisable customers.

You will probably establish a face-to-face engagement model with customers offering you a high return. And you will want to develop more of a self-help engagement model for customers with a low return.

You may also find the tools or approaches for your low-return segments useful in your high-return segments. Some of your larger customers will feel comfortable with self-help. With these customers, you will end up with a hybrid model of engagement.

The return from customers won't remain static. Regular review of your assessment of each customer will ensure you categorise each correctly.

Outcome Relationship

The second segmentation criterion involves the relationship customers want to have with you. Customers will respond differently to your outcome program. Some will immediately love your focus on their Success Outcome. And they will embrace working with you on that outcome.

You're likely to find other customers who don't buy into your outcome program at all. When they first bought from you, you both established a transactional engagement model. They don't want to change that. They'll say something like: 'We'll call you when we need to buy something. And by the way, when we do, we expect a big discount.'

Other customers may find your outcome program intriguing but feel uncertain. They will want to see other customers go first. Most will eventually join your program, but it may take time.

With customers who join your outcome program, you'll find different levels of maturity. Those with the highest maturity will probably become the best references. You're likely to invest more with these customers.

So, some customers will want an outcome engagement. And others will still want a transactional engagement.

Putting the Two Criteria Together

You can now allocate each customer to a matrix based on their economic return and their engagement preference.

It's not critical you follow this segmentation structure. But if you do, your segment model will look like this:

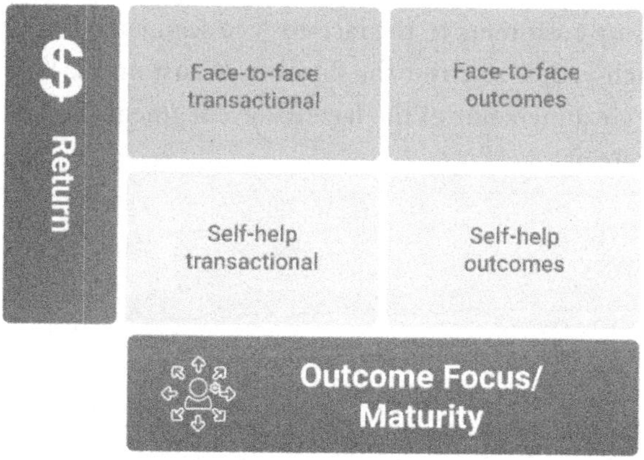

Figure 73: Customer segmentation model

The percentage of customers in each cell of the matrix will evolve. Initially you may have a minority of customers in the outcome cells. Over time, the percentage of customers embracing an outcome approach will increase.

In time, the majority will do so. But you'll probably always have some customers who want to remain with transactional engagement.

Different Industries

You may serve different industries. Each industry will use different language and will have a different PSPSPS model. If your business is large enough, you may also segment by industry. You can create a 2 x 2 matrix for each industry.

Allocating Resources Between Segments

You can now consider how much of your scarce resources to allocate to each of your segments. This exercise can create some interesting challenges. You will have in mind an ideal way of engaging with customers. But you can't always afford the ideal approach.

You may find compromise necessary. For example, you may have allocated many customers to the face-to-face segments. But you can't afford enough people to cover the number of customers. You need to move some customers out of the face-to-face segments and back to the self-help segments.

Section 2 – How You Engage With Each Segment

Let's consider what engagement in each of the four segments may look like.

Face-to-face Outcomes Segment

These high-return customers have embraced your outcome approach. You have established a repeating cycle of engagement with them. You focus, together, on continuous improvement of the Success Outcome. You jointly work to identify ways to improve the customer's business by improving the Success Outcome.

You've established a high degree of trust. The customer sees you as a trusted adviser, particularly your Success Consultants. They have faith in you.

You help customers improve their business results. So, you've established strong executive engagement. You have access to Deciders when you need it. And they welcome your engagement.

Customers regularly buy your products and services without considering competitive products. Your understanding of their business and your role in formulating business initiatives sets you apart. Buying from another vendor would introduce risk. Your customers feel other vendors don't have the same ability to enable outcomes.

Even if a competitor introduces a better product, you're unlikely to lose the customer. The customer understands they need more than products to achieve a business outcome. They have learned from you about the importance of getting all aspects of PSPSPS right. To displace you, a competitor would need more than a stronger product. They would need a stronger ability to enable the Success Outcome.

Self-help Outcomes Segment

Customers in this segment have also embraced your outcome approach. They have a similar attitude towards you as those in the Face-to-face Outcomes Segment.

A series of online tools, guides and methodologies leads your engagement. You will have methods for gathering outcome metrics with your customers. And you create practical, useful reports for them.

The customers use these reports to identify potential areas for improvement. They then use your other tools online to develop business initiatives. The focus of these initiatives remains the same—continuous improvement of the Success Outcome.

Face-to-face Transactional Segment

This segment contains high-return customers who still want an arm's-length relationship. They tend to see you as a product provider. They expect you to have strong expertise in your products. And to respond quickly if they encounter product problems.

If they seek advice, they will focus on improved use of the product. They don't see you as a business adviser.

In this segment you should aim to convert as many customers as possible to your Face-to-face Outcomes Segment.

Self-help Transactional Segment

This segment contains lower-return customers who still want an arm's-length relationship. For many vendors, this segment remains problematic. It's often your long tail. A group of customers that in total create meaningful revenue. But as individual customers, don't offer enough returns to warrant face-to-face engagement.

And because this group has not bought into your outcomes model, they are not strategic.

So, you will offer self-help focused on maximising use of your product. For these customers, provided you respond when they have a problem, this level of engagement may prove adequate.

Over time, you'd like more of these customers to move to the Self-Help Outcomes Segment. But it's not your highest priority.

Summary

Segmentation allows you to tailor engagement for different groups of customers. And to allocate resources in a strategic way.

You'll find customers who respond positively to your outcome program. You'll find others will resist and others who will feel uncertain. Over time, you should aim to move most customers to an outcome engagement. But it's likely you'll have some who don't ever convert.

PART 8 – IDEEAL Outcome Lifecycle for New Business (New Logo)

We introduced the concept of a Business Malady (Part 6 Section 4 of Book 2). It's a steppingstone to bring prospects into your outcome program. In Part 8, we will cover how to use the Business Malady in different New Business (New Logo) sales situations. We'll cover:

- Outbound Face-to-face
- Outbound Electronic Engagement
- Inbound Sales—resetting late-stage buying cycles

Section 1 – IDEEAL Lifecycle for Outbound Face-to-face

You may have people dedicated to an outbound process to find New Business (New Logo) prospects. They'll have titles such as Business Development Representatives, Sales Development Representatives or even Telemarketers.

Once this team has created contact with a prospect, you should aim to get the prospect into your BuILD process. You'll take the prospect through a series of steps, each of which involves a small extra commitment of time, effort and eventually money. You aim to help the customer commit to and then develop a plan for a business initiative. It's through the business initiative that you'll sell your products and services.

The following sets out an example of how you could run the IDEEAL Lifecycle using an outbound approach for New Business.

Investigation Starts

A buying cycle only starts when someone with authority (Decider) decides an aspect of the business needs to improve. They Decide to Investigate.

What Your IDEEAL Lifecycle Must Do	How to Do This in New Business (New Logo) Prospects—Outbound
Identify Deciders who've already decided to investigate	Contact the Decider. If it's your first contact, introduce the Business Malady and ask for 15 minutes to explore it.
Catalyse other Deciders to investigate	In your 15 minutes, explain how your Business Malady impacts similar companies to the prospect. Seek agreement that the Business Malady affects the

	Decider's business. Then, ask if they would like a high-level assessment of the impact on their business. Introduce your Business Malady Assessment. It's a short questionnaire (perhaps 10–15 questions). From the answers you provide a report on where the prospect is doing well, and where they might improve. (If possible, you should also develop a simple calculator for a rough-cut estimate of the financial impact of the Business Malady.) After getting answers to the short questionnaire, set a time to go through the report with the Decider. Write the Business Malady Assessment Report, then go through it with the Decider. The report should create the tension, desire or fear necessary to catalyse the Decider to act. The report also identifies the areas for improvement, where a business initiative would improve the business. You have completed the first two steps of the BuILD process—Investigate and Assess.

What to Do Defined

The prospect now needs a plan to achieve the better business results.

What Your IDEEAL Lifecycle Must	How to Do This in New Business (New Logo) Prospects—Outbound

Do	
Help the customer develop a plan for a business initiative Ensure you actively take part to build faith	Your Business Malady Assessment Report identifies where opportunities for improvement lie. Focus on these areas in building a plan for a business initiative. You may feel tempted to move straight to a proposal for your products and services. Avoid this temptation. The Decider doesn't yet have a clear plan for a business initiative. They won't have the confidence a clear plan will provide. The prospect is still in the Decide What to Do phase. So, phase-appropriate engagement means you can't talk about your products yet. Focus now on helping the prospect create a plan for a business initiative. And remain product agnostic. The 10% Sale provides an excellent way of creating the detailed plan the prospect needs. The study will create a detailed report with everything the prospect needs to get a business initiative approved. Some prospects won't pay for the 10% Sale in a New Business situation. You can ask them to 'pay' in other ways if you decide to proceed. For example, you may ask for access to executives. The detailed report will create the confidence the Decider needs to move to the next phase of their buying cycle. You have completed the final step of the BuILD process—Plan. You may also develop other methods for helping the customer achieve the confidence needed. You might experiment with a 'to-be state trial'. For simpler

	products, particularly where the product delivers the bulk of PSPSPS, you may quickly simulate the new to-be state. It's not a trial of your product. It's a trial of the new to-be state.

Engagement Won

To create revenue, the customer must select you as part of their business initiative. By this stage, you have done a great job ensuring the customer focuses on everything required (PSPSPS) to achieve the business results.

What Your IDEEAL Lifecycle Must Do	How to Do This in New Business (New Logo) Prospects—Outbound
Ensure selection focuses on ability to enable the Success Outcome	The faith built in the first two steps may mean the prospect does not consider competitive vendors. You have proven you know the initiative and will focus on their results. If there is a competitive evaluation, you can use the detailed report from the study. It addresses everything needed to achieve the business results (PSPSPS). You can help the customer extract the headings as decision criteria.
Ensure final approval occurs	Before writing the detailed report, you enquired about the format required in a submission for approval. You ensure the report covers everything needed for approval. In the format the customer needs. The report becomes the submission for approval.

The Rest of the IDEEAL Lifecycle

The prospect has now become a customer.

You will take them through the remaining three steps of the lifecycle:

- Outcome Enabled, Measured
- Advocates Created
- Cycle Looped

Putting the six steps together for New Business prospects using an outbound approach looks like this:

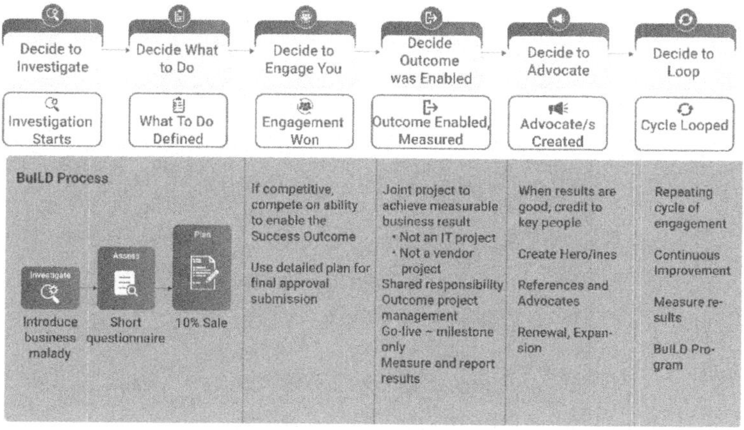

Figure 74: IDEEAL lifecycle for the Outbound Face-to-Face segment

Section 2 – The IDEEAL Lifecycle for Electronic Engagement

You'll need to adapt the IDEEAL Lifecycle for New Business prospects when your early engagement is electronic.

Investigation Starts

What Your IDEEAL Lifecycle Must Do	How to Do This in New Business Prospects—Electronic Engagement
Identify Deciders who've already decided to investigate	Deciders Already Decided to Investigate To identify Deciders who have already decided to investigate, prompt them to engage with you electronically. They have already experienced tension, desire or fear and decided they need to do something. They're about to move into their Decide What to Do phase. They will need to build a plan for a business initiative. If you provide help with building that plan, they may engage you. We'll offer some suggestions on how you might help them in the What to Do Defined step. You can alert Deciders to your ability to help them via: • Emails to them • Emails to staff in the customer who could advise

	the Decider • Online content • Content in publications they read • Hard-copy communication—letters, pamphlets
Catalyse other Deciders to act	Catalysing Other Deciders to Act The Decider needs to feel tension, desire or fear before they will act. You can use either or both of two approaches. In the first approach, use your Business Malady like this: • Electronically introduce the prospect to the Business Malady. You can use the same methods as set out above, such as emails, content, etc. • Provide an online tool for them to answer the 10–15 questions in your Business Malady Assessment. Use an automated tool to create a custom report for the prospect. The report should highlight areas the prospect is doing well and areas to improve. Ideally, include a simple calculator that will do a rough-cut cost of the Business Malady. The second approach uses your Success Outcome. Focus on the to-be state the prospect could achieve. Methods you might use to create the tension, desire or fear include: • If you have access to their operating data, create a report that highlights current performance in the Success Outcome. Show trends over time. If possible, provide benchmarking data that compares them to industry standards.

	• Provide insights into activities or performance of their competitors in the Success Outcome. • Alert them to any external triggers such as regulatory changes. • Run webinars discussing industry performance in the Success Outcome. • Provide an online tool for them to self-assess their performance. By answering a limited number of questions, you'll give them insight into how they are performing. This will be the same, or very similar, to your Business Malady Assessment (but with a Success Outcome title). Over time, you'll see which of the two approaches (Business Malady or Success Outcome) produces the strongest flow of leads.

What to Do Defined

What Your IDEEAL Lifecycle Must Do	How to Do This With Electronic Engagement Prospects
Help the customer develop a plan for a business initiative Ensure you	Develop one-to-many approaches for helping prospects build a plan for a business initiative. Ensure they consider everything required to achieve the better business results (PSPSPS). As part of the previous step, the prospect may have answered the 10–15 questions of your online

actively take part to build trust	questionnaire. Or they attended a high-level webinar. They should have a high-level understanding of a path to success. To complete the confidence they need, they now need a detailed plan. To help them develop a detailed plan, you can offer a few options: • Provide a second-step online questionnaire with more questions (more than 30 and less than 100, depending on the complexity of your products). Using an automated tool, create a plan. • Run regular webinars on building a plan. • Offer a video-based guide and a template for building a plan. • Offer your 10% Sale. See below for more detail. By the end of this process, the prospect will have a detailed plan for a business initiative with a full business case. White papers offer a more traditional approach to guiding prospects. While they may have a place, their passive nature makes them less effective. You want to actively assist in developing the prospect's plan. Even if it's only through electronic means. 10% Sale For prospects where you would engage face to face, you may use the 10% Sale technique. Explain how you will jointly do a detailed study with the prospect. Be product agnostic. And create a detailed report with everything the prospect will need to get a business

	initiative approved. To help the prospect understand the nature of the report, provide a downloadable template or a redacted version of a real report. You would normally expect a customer to pay for your role in the 10% Sale. With prospects, this may prove more challenging. You might seek 'payment' in other ways. For example, you'll conduct the study if you have access to the right executives and they agree to give you the necessary time.

Engagement Won

What Your IDEEAL Lifecycle Must Do	How to Do This in New Business Prospects—Electronic Engagement
Ensure selection focuses on ability to improve the Success Outcome	Now, you need the prospect to select you as the vendor they will work with. Even though engagement has been electronic, you've helped the prospect through your BuILD process. All without discussing your products. You've built faith. The faith will stand you in good stead during an evaluation of vendors. As part of the engagement, you have also set expectations about what it takes to achieve the better business results. You've discussed PSPSPS. You now want to ensure the prospect's selection criteria focus on PSPSPS. You don't want their selection process to descend into a feature/function

	shootout. The good news is the detailed plan the prospect prepared has all the headings needed to guide their selection criteria. You just need to help them focus on the right headings.
Ensure final approval occurs	Once they have selected you, you need the prospect to get final approval for the business initiative. Smaller prospects may not have a formal approval process. The Decider will often be the Approver. These customers may not develop a formal submission. After negotiating the deal, you may have done enough to close the deal. Most prospects will have a formal approval process. The detailed plan you helped them develop provides the foundation for their internal submission. You've already done most of the work.

The Rest of the IDEEAL Lifecycle

The prospect has now become a customer.

You will take them through the remaining three steps of the lifecycle:

- Outcome Enabled, Measured
- Advocates Created
- Cycle Looped

Putting the six steps together for New Business—Electronic engagement looks like this:

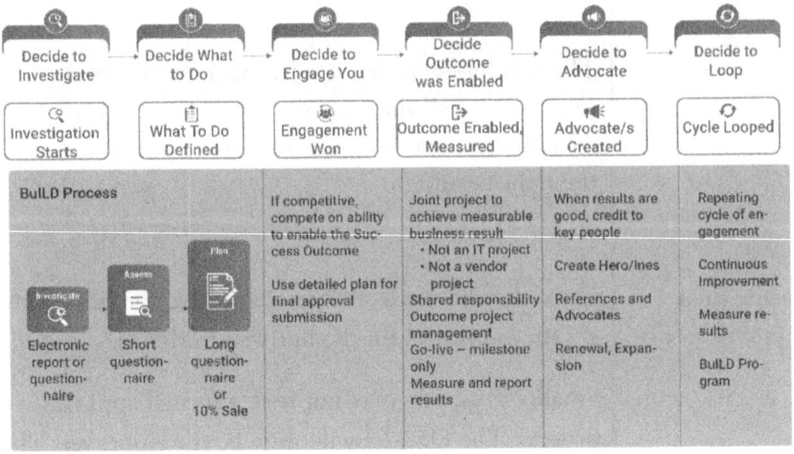

Figure 75: IDEEAL Lifecycle for electronic engagement

Section 3 – Inbound Sales – Resetting Late-stage Buying Cycles

In Sections 1 and 2, we looked at outbound and electronic New Business (New Logo) opportunities. Now let's consider inbound opportunities.

A customer or prospect has contacted you. They tell you they have selected you as one of several vendors to evaluate. They want to arrange presentations and demonstrations to learn about your offering.

The situation has all sorts of potential problems. First, it's possible another vendor has influenced their selection criteria. In that case, you face loaded dice. Winning will prove challenging.

Second, even if the dice aren't loaded, you'll have no more chance of winning than any other vendor. The prospect will focus on the products from each vendor. It will turn into a feature/function shootout. Which anyone can win.

Third, the prospect may not have a well-defined business initiative. They have a problem and have decided technology will fix it. But that's the extent of their thinking. They haven't thought through what they need to achieve.

Once they learn more, they may realise technology isn't the answer. And they will decide to do nothing. Vendors have a major headache with customers or prospects entering evaluations and later deciding to do nothing. Do-nothing decisions often proves to be the biggest 'competitor'.

Fourth, you may win. But because they don't have a well-defined business initiative, the project loses direction. They know they want a problem fixed, but they don't know how they want things to be. In the end, they regard the project as a failure. And that will cost you future revenue and impact your reputation.

Resetting Their Buying Process

A salesperson's ability to influence a late-stage buying cycle is a bit like a parent's ability to influence a teenager. Not much. So, your salespeople need a different approach. You will attempt to reset the prospect's buying process and take them back into your BuILD process.

If you're successful, you'll help your chances of winning. But you'll also help the prospect. Without a well-defined business initiative, they will find it hard to:

- choose the best vendor to deliver improved business results. They'll decide on who has the best feature/functions. And that's only a small part of what's needed for success.
- create a compelling business case. If they are not clear about the to-be state they are trying to create, how will they quantify the benefits? The lack of a compelling business case reduces the likelihood of approval for the project they need.
- achieve success. A poorly defined project has a low chance of success.

By taking them back into your BuILD process, you will help them develop a clearly defined business initiative. You will help them understand everything required to achieve success. And that's in their interest.

Using Your Business Malady

To disrupt a late-stage buying cycle, here are the key steps:

- Listen to the problems the customer has in mind.
- Tell them you have many other customers who faced similar problems.
- Tell them the problems are symptoms of a Business Malady.
- Introduce your Business Malady and discuss its applicability to the prospect.
- Offer your Business Malady Assessment and create a custom report. Show the prospect where they are doing well and where they could improve. Assess the financial impact of the Business Malady.

- Help them develop a detailed plan for a business initiative.
- Influence the selection criteria to focus on everything needed to achieve the Success Outcome—PSPSPS.
- Once you're selected, use the detailed plan as the submission for final approval.

Section 4 – What If a Prospect Will Not Focus on Outcomes?

If you have a prospect who won't focus on an outcome approach, you have a policy decision to make. Your competitive advantage lies in selling your Success Outcome and your ability to enable it. If the customer won't engage on your Success Outcome, you're back to competing on your products and services alone.

What is your policy when this happens? It's incredibly difficult for salespeople to walk away from a prospect. Where there's a deal, there's always hope. Surely you should try to win it.

Two factors might cause you to re-evaluate that thinking.

First, if you offer subscription pricing, you can lose money even when you win a deal. You should have some idea of your customer acquisition cost (CAC), i.e. how much it costs you to win a deal. You will also know the revenue from a deal in the first contract period. If the CAC is higher than first-period revenue, and the customer does not renew, you have lost money.

For many vendors, it's a new phenomenon.

The second factor that might influence you to focus on outcome-based deals revolves around win-rates. If you sell to any prospect, you will have a certain win-rate. If you sell only to prospects who embrace your outcome approach, you will increase your win-rate.

Higher win-rates will reduce the average cost per sale and increase the revenue per salesperson. And both will have a positive impact on your bottom-line.

Summary

New Business (New Logo) prospects won't be familiar with your outcome approach. You can use your Business Malady as a steppingstone.

The Business Malady Approach supports your outbound program and electronic engagement.

For inbound prospects late in their buying cycle, you'll attempt to disrupt their process. You'll bring them back into your BuILD process.

For prospects who won't engage with you on outcomes, you have a decision to make—do you continue or not?

PART 9 – Outcome Measures

You can use three different measures of outcomes as part of your outcome program.

Customer Outcome Achievement

Jointly with the customer, you will measure the Success Outcome and Contributing Outcomes. Together, you will calculate the benefits of improvements in these outcomes. You can use these measures as a high-level guide for the next round of improvements in the outcomes.

Vendor Execution Scores

You will track how well you're executing your IDEEAL Lifecycle. You'll develop measures for each of the six steps.

Success Likelihood Score

You can use this to assess the likelihood of success with a prospect or customer. You can use it to guide you on whether to engage with individual prospects or customers. In the case of customers, you may choose to disengage. If not, you may seek changes from the customer to increase the likelihood of success.

Section 1 – Customer Outcome Achievement

Enabling a business outcome creates a new to-be state for the customer. It establishes a new level of performance that higher management regard as a success.

The customer will want to maintain and improve this new level of performance over time. And that will help you keep the customer and drive expansion revenue.

You need to measure that new level of performance.

Don't Measure the Value of Your Products

You're not trying to measure the value of your products. You're measuring the value of improvements in the Success Outcome. As long as the customer sees you as key to the Success Outcome, they will keep you. And grow their engagement with you. There's no need to ask them to assign a specific value to your products.

This change of measurement reflects the change in your engagement with customers. You now have a collaborative approach. At every step of the IDEEAL Lifecycle, you work jointly to improve the Success Outcome.

Align Measurement with Your Decider's Success

Before any buying cycle commences, a Decider must Decide to Investigate. Your core measures should help the Decider Decide to Investigate.

If you think back to the Outcome Hierarchy, each role will have a core outcome they deliver for their company. You've identified your Decider. You have aligned your Success Outcome to the core outcome your Decider owns. You play a key role in helping the Decider deliver their core outcome.

You are the most significant external provider affecting the Decider's success.

You should understand how your Decider is measured. And align the measurement of your Success Outcome to how the Decider's success is judged.

Let's look at a couple of examples.

ERP Vendor

Companies buy ERP software to help their operations run effectively. A customer's COO will have effective operations as a key measure of success. So, an ERP vendor may identify the COO as their Decider—the key person with whom they will align.

The COO will have measures of success that might include gross margin, revenue per person or cost per unit of production. If you can find a single measure the COO regards as most important, focus on that. Otherwise, you might agree on a set of measures.

Marketing Automation Vendor

Companies invest in marketing automation not just to increase pipeline, but to increase sales. So, marketing must create Closable Pipeline. If not, the CEO will consider the marketing automation system a failure.

Typically, the CMO owns pipeline. If a vendor's marketing automation offering addresses all aspects of pipeline creation, they may align with the CMO. The CMO is their Decider.

Examples of Closable Pipeline measures include:

- Increase in revenue from marketing-created leads
- Percentage of sales revenue from marketing-created leads
- Marketing cost per dollar of revenue from marketing-created leads

These measures show how the marketing automation vendor has helped the customer's business. It's a collaboration between the vendor and the marketing department. Which is why the vendor shouldn't try to measure the value of their products. Closable Pipeline results from the joint work of vendor and customer.

Three Business Imperatives

You should link Customer Outcome Achievement to one of three business imperatives:

- Increased revenue
- Decreased costs
- Reduced business risk

If you don't link what you do to one of these three imperatives, the customer may not perceive the full value of your relationship.

You may feel the outcome you help customers achieve has a more indirect value. For example, if you have a Human Resources offering, you may aim to improve employee experience. Measuring employee experience might prove difficult. But an improvement in employee experience may increase retention. Reduced turnover will reduce recruitment costs.

Becoming Indispensable

Hopefully, you will become integral to the way a customer works. The customer can't envisage working without your products and services. In this case, the customer may not feel the need to measure the value you help them achieve.

While not going to the trouble of measuring value may be tempting, it has dangers. If something changes, you may be exposed. For example, your Decider might leave. A new Decider might want to revisit their predecessor's approach. If you don't have a measurable track record, the new Decider might conclude you are not indispensable.

Section 2 – Vendor Execution (How Well You're Executing Your IDEEAL Lifecycle)

We've discussed the six steps of the IDEEAL Lifecycle, the six IDEEAL Decisions and the six emotions required:

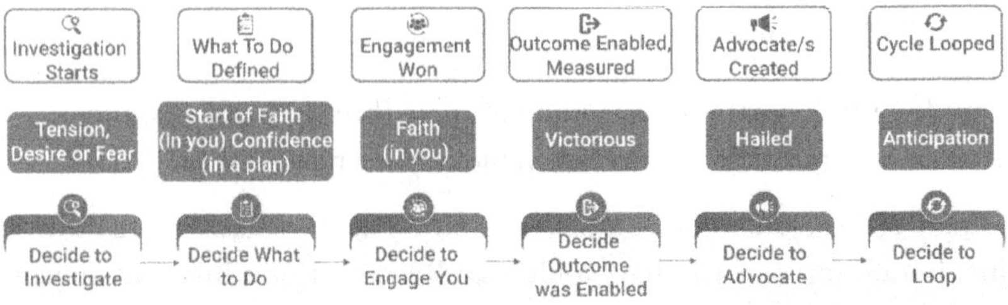

Figure 76: The six steps, emotions and decisions

You'll want to know how well you execute each step. In this section, we will discuss the measures you might put in place.

Ideally, you'll choose only a few measures for each of the six steps.

Most will be lag indicators of activities completed. But some of these will also act as lead indicators for later steps. For example, in your Executive Business Reviews you aim to identify possible business initiatives. The size of the pipeline for possible business initiatives will indicate the number of future deals you might win.

What are some possible measures for each step of the IDEEAL Lifecycle?

Investigation Starts

In this step, you want Deciders to Decide to Investigate. You should track how many Deciders make this decision.

We discussed earlier the two states for Deciders:

Already Decided to Investigate

You want to identify Deciders who have already decided to investigate. Something has already caused them to feel tension, desire or fear, and they are moving forward. They are in the Decide What to Do phase.

Unless they reach out to you in some way, it's hard to track this. You need to offer them something that will help them Decide What to Do. That way, they will reach out to you.

One method involves creating tools, guides, methodologies. Let all Deciders know they can access the tools, guides, methodologies on your website. After Deciders Decide to Investigate, they will visit your website. Initially they may browse. But you can offer something that requires them to enter contact details.

You may also track the number of inbound calls from Deciders or people acting on behalf of Deciders. Make it easy for Deciders to call you.

You now have an idea of the number of Deciders who are in the Decide What to Do phase.

Not Yet Decided to Investigate

You want to catalyse these other Deciders to investigate.

For face-to-face customers, tracking the output from EBRs provides a good guide. You can track the percentage of EBRs which result in you moving to the Assess step in your BuILD process.

For self-help customers, you can offer the same tools, guides and methodologies for their use. And you can track the number of self-help Deciders who utilise them.

Measures

Measures for this step might include:

- Number or percentage of companies where Deciders move into the Decide What to Do phase

You might segment based on the numbers from:

- New Business (New Logo) prospects
- Existing customers using self-help engagement
- Existing face-to-face customers

What to Do Defined

Prospects and customers go through two steps in the Decide What to Do phase.

First, in the Assess step of BuILD, they develop a high-level plan for how they will achieve the better business results. For face-to-face customers, you will work on the Assess step. You can track the value of possible business initiatives that have a positive business case.

For self-help customers, you can track usage of your online tools. Again, you can measure the value of possible business initiatives with a positive business case. This time, the customer has used your online tools to model the possible projects.

The second step in this phase is the Plan step of BuILD—development of a detailed plan for a business initiative. You can now track the value of possible business initiatives arising from the detailed reports you create with the customer. You should also track the conversion rate, i.e. what percentage of detailed reports resulted in a business initiative proceeding?

Measures

High-level Plan (Assess) step

- Face to face
  - Number and value of possible business initiatives with a positive business case
- Self-help
  - Number of companies using your self-help tools

Detailed Plan (Plan) step

- Face to face
  - Number of detailed studies undertaken

    - Pipeline arising from those studies
    - Conversion rate (how many studies result in deals)
- Self-help
    - Pipeline arising from this step
    - Conversion rate

Engagement Won

You will track the value and conversion rate for different sources of leads, with focus on leads from your BuILD process.

Measures

- Win-rates by segment and source of lead

Outcome Enabled, Measured

In this step, you should measure how well you're executing your outcome-based project management approach.

Measures

Measures might include:

- Project setup—the degree to which you establish the 'ideal project'. For example, were measurable business objectives established? Was the right governance set up?
- Project execution—the degree to which you followed the ideal process, i.e. in practice was your outcome methodology followed?
- Measure and report—were results measured and reported back to the Decider?
- Results achieved—to what extent did the project deliver the project objectives?

Advocates Created

This step focuses on creating references and advocates. You will want to track the percentage of projects which resulted in advocates. But it's not easy to know whether someone is acting as an advocate. Tracking mentions on social media can help.

Or, you may measure the number of times you exercised your hero/heroine program. In other words, you ensured your champions in the customer received credit for the successful project.

Measures

- Number of companies where you ran your hero/heroine promotion
- Number of reference accounts
- Number of advocates (where you can detect people acting as advocates)

Cycle Looped

For this step, you're interested in the percentage of customers where you have established a repeating cycle of engagement. You will have agreed how and when measurement of the Success Outcome and Contributing Outcomes will take place. You'll have a process for reviewing the results. And identifying the next opportunities for improvement.

Ongoing Measurement

In step six of the IDEEAL Lifecycle (Cycle Looped), you agree on a repeating cycle of engagement with the customer. You should include a plan for ongoing measurement of the Success Outcome and Contributing Outcomes.

Your plan should include:

- The measures to track
- How often you will gather the measures
- Who will gather the data
- Who will create reports
- How you will review the reports

The measurement of the Success Outcome and Contributing Outcomes underpins the repeating cycle of engagement with your customers.

Measures

- Number or percentage of customers agreeing to a repeating cycle of engagement

IDEEAL Measurement

You should track the participation by customers in your outcome program. You'll want to know how many of your customers have embraced your outcome approach. You should also know the percentage of total expansion revenue created by your outcome program.

In both cases, the numbers will start small. But then grow over time.

Measures

- Number or percentage of customers engaged in your outcome program
- Percentage of expansion revenue from your outcome program

Section 3 – Success Likelihood Score

Successful Customers Are More Important Than Ever

Subscription pricing has moved the power back to customers. It's lowered the barriers to entry and exit. Now, if customers don't feel successful, they can cancel, reduce or stagnate their subscription. That will cost you future revenue.

You can lose money making a sale. The customer acquisition cost (CAC) may be higher than the revenue in the first contract period. If the customer cancels at the end of the first period, you have lost money.

Losing customers also undermines staff confidence. And that increases staff turnover and reduces discretionary effort. Both will cost you money.

Finally, with the availability of information online, the loss of a customer damages your reputation more than ever. And that might cost you future revenue.

The financial imperative for having successful customers has never been stronger. And that means choosing the right customers has never been more important.

In this section, we will provide a framework for reviewing the likely success of customers.

Three Views of Success

Business Fit

Are you and this customer or prospect a good fit for each other? What's the likelihood of long-term success?

Customers who aren't successful over time become a burden. They absorb your resources with little or no return. And they are likely to become negative references.

If a relationship proves problematic, the Business Fit score can help you see why. It may inform corrective action you can take. And in extreme cases may prompt you to disengage entirely.

Project Success

You drive expansion revenue through business initiatives. You execute each business initiative through a project. Before you start a project, you will want to understand the likelihood of success. And whether you need to take any corrective action.

Expansion

How likely is the customer to drive improvement in the Success Outcome? Will they offer opportunities for expansion revenue?

Customers offering little or no expansion revenue make it difficult to achieve growth targets. For example, you may have a plan to grow current revenue by 20%. Consider what it means if, say, half of your customers are stagnant, i.e. no expansion revenue. To achieve 20% growth overall, the other half of your customers will need to grow at 40%. For most vendors, that's challenging.

You'll find a detailed guide to calculating the Success Likelihood Score in Appendix 3.

Redseal helps organisations achieve cyber resiliency. Ramesh Kaza, Vice President of Customer Success said:

Our conversations with customers focus on their cyber resiliency. To help customers, we have developed a Digital Resiliency Score. It's a single number that tells customers the status of their cyber protection and resiliency.

The Digital Resiliency Score includes assessments of the key elements of cyber resiliency—visibility, compliance and risk management. We assess customers on each element. We then apply an algorithm that weights the different factors affecting resiliency. The result is a score out of 850.

We use this score in a few ways. First, it underpins our quarterly business reviews with customers. The score gives an overall impact of customers' progress on their cybersecurity initiatives. We can then break the score down into its elements. And that helps customers identify actions they may need to take.

We also produce a cyber resiliency executive summary report with other key risk indicators. Senior management find this particularly helpful.

Finally, we make the reports and a dashboard available on-demand. So, any time a customer needs further insights about their cyber resiliency, they can get the most up-to-date facts about their situation.

The Digital Resiliency Score has changed the way we engage with customers. We don't need to talk about our products, or adoption, or key feature usage. The conversation focuses on the business outcome of cyber resiliency. The actions customers take to maintain or improve their resiliency drives our expansion revenue.

Everyone wins.

PART 10 – Evolution of Customer Engagement

There's been an evolution of customer engagement underway for decades. Vendors have evolved from

- A feature/function focus, to
- Solution selling, to
- Value selling, and in more recent years to
- An outcome-based approach.

It's a journey. It takes time and commitment to move from one generation to another. But the industry has shown it will evolve. The biggest shift occurred in the late 1980s and early 1990s with the advent of solution selling. Before that, vendors concentrated on demonstrating the features and functions of their products. Winning deals became more difficult as products matured and differentiation became more difficult.

So, solution selling emerged. Identifying customer problems and showing how the vendor provided a solution to those problems. Solution selling has served the industry well. But the industry is moving on.

Emergence of an Outcome Approach

To address the challenges with traditional engagement, outcome-based engagement has emerged.

The following chart intends to portray that the strength of customer relationship increases as you move through different phases of customer engagement.

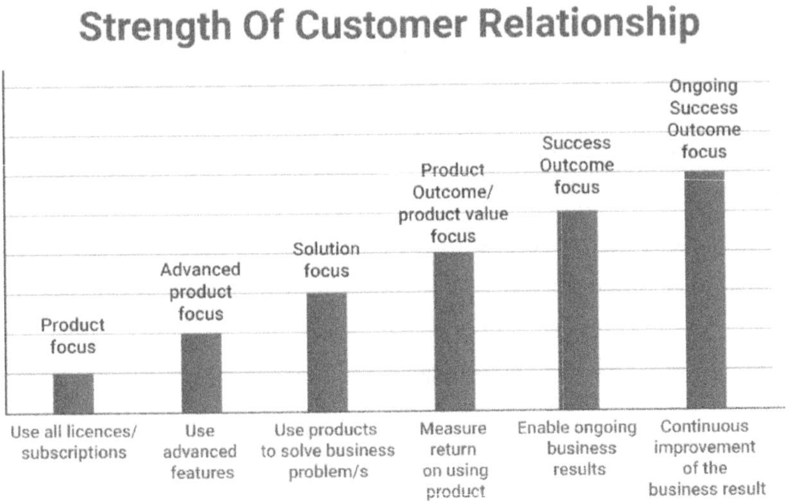

Figure 77: Strength of customer relationship with different engagement models

In the following table, we will set out the characteristics of engagement of the different phases of engagement.

	Product Focus	Advanced Product Focus	Solution Focus	Product Outcome/ Product Value Focus	Success Outcome Focus	Income from Outcomes Focus
Nature of Customer Engagement	Transactional. Vendor push. Intermittent engagement about the product and its use	Transactional. Vendor push. Stronger engagement about the product and its use	Transactional. Vendor push. Identify problems the customer may have that lead to use of the product	Transactional. Vendor push. Vendor attempts to quantify return from using the products	Collaborative but still transactional. Jointly evaluate business initiatives to improve the Success Outcome	Collaborative and repeating. Agreed cycle of engagement to jointly evaluate business initiatives to improve the Success Outcome
Who We Engage With	Intermittent middle management, technical recommenders (IT), users	Intermittent middle management, technical recommenders (IT), users	Middle management, technical recommenders (IT), users	Intermittent executive engagement, middle management, technical recommenders (IT), users	Executive engagement, middle management, intermittent with technical recommenders (IT), intermittent with users	Recurring executive engagement, middle management, intermittent with technical recommenders (IT), intermittent with users

Customer Executive Perspective	Not sure what benefit we get, but the users like it	Not sure what benefit we get, but the users really like it	Not sure how to quantify the benefit, but it helped us solve a problem	This product has a positive ROI	This vendor helped us achieve a new level of business performance (that I'm measured on)	This vendor helps us every year to improve the performance (I'm measured on)
Marketing Messages	We have a great product, these customers like our product	We have a great product, these customers like our advanced features	Our product solves these problems, and these customers have solved them with our product	Here's the value customers get by solving problems with our products	Here's the Success Outcome we enable, here's why it's important and why we're best at enabling it	Here's how we improve the Success Outcome over time
Sales (to Existing Customers)	Show the benefits of the product	Show the benefits of the product's advanced features	Find a problem, show how the product solves it. One sale at a time	Find a problem, show how the product solves it. One sale at a time	Introduce the Success Outcome, jointly evaluate business initiatives to improve it	Work on a pipeline of possible initiatives that can improve the Success Outcome
Sales—New Logo	Show the benefits of the product. Use references to attest to how great the product is	Show the benefits of the product's advanced features. Use references to attest to how great the product is	Identify problems, show how the product solves them. Use references to show how these problems have previously been solved	Identify problems, show how the product solves them. Provide an ROI. Use references to show the value other customers have achieved	Acknowledge existing problems. Introduce the Success Outcome, show how it solves the problems. Show expertise in everything required to achieve the Success Outcome. Use references to show how to achieve the Success Outcome and the benefit	As with outcomes but show the continuous improvement model. Use references to show how to improve the Success Outcome over time

Pre-sales Consulting	Show the features of the product. Free-of-charge work	Show how to use the product to solve problems. Free-of-charge work	Show how to use the product to solve problems. Free-of-charge work	Show how to use the product to solve problems and how to measure benefits. Free-of-charge work	New role—Success Consultant. Customer and Success Consultant jointly analyse business initiatives to improve the Success Outcome. Customer pays for some of their work. Product pre-sales still play a role—prove the product can support the business initiative	Success Consultants key in the continuous improvement process. Customer pays for some of their work
Success Consulting	NA	NA	NA	NA	Guide the customer on all aspects of achieving the business outcome (including 'non-vendor' elements). Take part in implementation project	Lead an ongoing process of improving the Success Outcome
Onboarding (Simpler Products)	Get users onto the product	Get users on the more advanced features	Get users on the more advanced features	Get users on the more advanced features	Get users on the more advanced features	Get users on the more advanced features

Onboarding And Professional Services (More Complex Products)	Run a project to get the product live	Run a project to get the product live, including advanced features	Run a project to solve a business issue with the product	Run a project to solve a business issue and obtain agreement on the benefit gained by the customer	Run a project to enable a business outcome (which may solve a business problem)	Run an ongoing series of projects to continuously improve the Success Outcome
Customer Success	Adoption—use all subscriptions	Adoption and use the product's advanced features	Adoption and use the product's advanced features	Measure the benefit from using the product	Initiate business reviews, measure KPIs for the Success Outcome. Success Consultants to help	Set up, monitor and drive the ongoing joint engagement
Support	Fix product problems, reactive	Fix product problems, reactive	Fix product problems, reactive	Fix product problems, proactive	Fix product problems, identify possible threats to the Success Outcome	Fix product problems, identify possible threats to the Success Outcome over time

PART 11 – Outcomes, Value and Pricing

In Part 11, we will consider the relationship between these three topics.

We'll offer an observation on what it takes to get a buying cycle happening.

We'll then talk about the evolution of pricing, where it may be heading, and what that means for vendors.

Section 1 – The Relationship Between Outcomes, Value and Pricing

Here is a proposition:

- An outcome creates a new to-be state.
- Value is what the new to-be state is worth to the customer.
- Price is what the customer will pay for your help to achieve the to-be state.

Let's explore this proposition.

Outcomes

We discussed earlier that whenever there's a commercial transaction, there's an outcome in play. Think hamburgers, cinema tickets and sports cars. So, whether it's clearly defined or not, if money changes hands, the payer expects an outcome. They expect things to be different and better.

In B2C, an outcome can be temporary. If you eat a hamburger, you become full. That's the new to-be state. But at some point you will feel hungry again.

In B2B, customers don't expect an outcome to be temporary. They expect a new level of performance in their business. And that performance should be maintained and improved over time.

An outcome creates a new to-be state for a customer.

Figure 78: An outcome creates a new to-be state for a customer

Value

Customers want to achieve the new to-be state. Value is what that to-be state is worth to them.

Three things affect people's view of the possible value:

- What's In It For Me (WIIFM)—no one initiates a project unless there's something in it for them personally.
- What's In It For the Company (WIIFC)—is there a suitable return for the company?
- Risk—how much risk do I bear if I support this project?

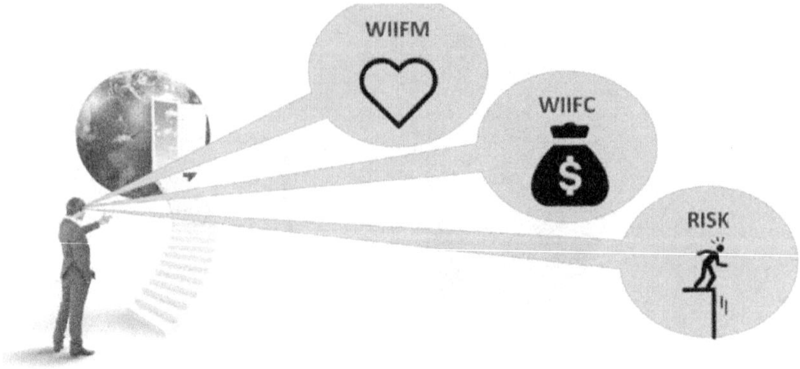

Figure 79: Three things affect perception of value

Let's imagine value as a pie. The balance of these three elements determines the size of the Value Pie.

Figure 80: Value Pie

Price

People will pay to achieve the new to-be state. The amount they will pay you depends on:

- The value they perceive
- How big your role is in enabling the to-be state
- How the financial risk of the project is shared

In effect, you will receive a slice of the value pie. The price will be high if:

- The value pie you enable is large
- You play a major role
- The financial risk for the customer is balanced

If the reverse is true, your price will be low.

So, what you do when you set a price is make a judgement on:

- The value of the to-be state you enable
- How much of that value a customer will give you

You come up with a price and then sell at that number.

Maximising Your Revenue From a Deal

To maximise the size of your slice of the value pie, you can do three things.

Make the Pie Bigger

You can make the value pie seem bigger. Use the Success Outcome to determine value. Don't focus on the value of your products (the Product Outcome). Focus on the value of the new to-be state (the Success Outcome). The higher the perceived value, the larger your slice will be.

Figure 81: Make the pie bigger

Do More to Bake the Pie

Perhaps this is what you currently offer:

Figure 82: What you offer

But creating the value pie takes a lot more. The customer needs:

Figure 83: What the customer needs for the value pie

You can get a bigger share of the value pie by offering more of what the customer needs to make the value pie. Achieving a business outcome requires PSPSPS. The more of this you deliver, the more of the value will be credited to your efforts. And the more the customer will pay.

Make Baking the Pie Lower Risk

You can minimise the risk for the Decider:

- Show a track record of enabling the Success Outcome.
- Develop a clear and complete plan to create confidence.
- Establish joint responsibility and accountability.
- Change your pricing to accept greater risk.

Value Is an Individual Judgement

Emotion and circumstances affect the perception of value. For example, imagine you have a bottle of water to sell. One potential customer is in a desert, dying of thirst. Your bottle of water will save their life. Another potential customer is in an oasis, surrounded by luxury and with ample water.

The value of the bottle of water is clearly different for these two people.

In B2B, more than one person will play a role in a buying decision. Each person will perceive different value. And that's because both WIIFM and risk varies by individual. One person might expect project success will bring a promotion or a bonus. Another might feel they won't benefit, someone else will get the glory. As with the bottle of water example, the value is different for those two people.

People will also consider how much personal risk they face. Risk balances WIIFM. If personal risk is higher than WIIFM, they will oppose the project.

Section 2 – Selling Value vs Selling Outcomes

We have discussed that decision-making, including buying, begins with an emotional commitment. People then rationalise the emotional decision. The flow in buying is WIIFM followed by WIIFC. In other words, people first consider how they might benefit, then consider how the company would benefit.

Value-based selling typically emphasises how much a customer will benefit if they purchase a vendor's products. The primary focus is WIIFC.

If you lead with WIIFC, you run the risk that a buyer's emotions will not be invoked. Without the right emotion, there's no buying cycle.

A Success Outcome Acts Like a Travel Poster

Your Success Outcome will create a mental image in the mind of an executive. And the executive should react with, 'I want to be in the picture!' You have invoked emotion. WIIFM has been activated.

You therefore have a much greater chance of a buying cycle proceeding. Of course, WIIFC will need to be established for a deal to happen.

The proposition, therefore, is that you should lead with an emotion-invoking to-be state, then work on WIIFC.

Section 3 – The Evolution of Pricing Models

Pricing Models

Let's consider a pricing maturity model with four pricing models. As you progress, you increase your collaboration with the customer and reduce the customer's risk.[(4)]

Pay for Access Upfront

The first level of pricing maturity involves the customer paying for full access upfront. Software perpetual licensing provides an example. Customers pay for the licences before they use them. The customer must license the maximum number of users needed, even if maximum usage occurs only intermittently.

All the risk lies with the customer. If they don't achieve the to-be state, the customer has little recourse. They feel locked in as they have bought everything upfront.

Pay for Access Over Time

This is the software subscription pricing model. Customers still need to license the maximum number of users they believe they will need. But payment is spread over time.

In this model, the risk has moved back to the vendor. If the customer doesn't feel successful for any reason, they can cancel or reduce their subscription. Now the vendor has a financial incentive to ensure the customer feels successful.

Note that most subscription models result in a higher cost to the customer over time. But the other advantages outweigh this higher cost.

Consumption

This is the model used by companies such as AWS (Amazon Web Services). Instead of paying for maximum usage, customers pay for the capacity consumed at any point in time.

In this model, the risk lies squarely with the vendor. If the customer consumes no capacity, there's nothing to pay. Again, the vendor has a financial imperative to ensure the customer feels successful.

This approach feels more like a partnership. Both parties have interest in ensuring the to-be state is achieved.

Pay-for-Results (Outcomes)

This is the model described earlier in the section on servitisation. The vendor provides products for free, and charges for the results achieved by the customer. Rolls Royce charge for the number of miles their engines fly. Phillips Lighting charges Schipol Airport for the light delivered, not the number of light bulbs used. Baxi Boiler Systems contract for an agreed number of 'warm hours' regardless of how much energy is used. Profitwell charges customers on outcomes achieved not for the software.

This model has very low risk for the customer. And in most cases the vendor plays a much bigger role in enabling the to-be state.

Impact on the Amount the Customer Will Pay

As you move through the maturity levels, your relationship with customers changes. The amount a customer will pay increases.

In the software industry, consider the move from Pay for Access, Upfront (perpetual licensing) to Pay for Access Over Time (subscription). Customers know they will pay more over time. But embrace it anyway. One of the reasons is the change in the risk balance. Subscription shifts risk from the customer to the vendor.

Vendors have benefited enormously. The valuation multiples on offer for subscription-based companies far exceeds those for perpetual or upfront payment companies.

Now consider the move to consumption-based pricing. The risk lies even more with the vendor. And yet AWS booms.

The final step is a Pay-for-results (outcomes) model.

Profitwell has conducted a deep analysis of the impact of this model. Here's what they found.

The Impact of an Outcome Approach

Profitwell carried out a study involving 3,800 subscription companies in different industries (Profitwell, 2018). They considered the impact of different pricing models on churn and on expansion revenue.

They first compared companies using feature-based pricing to those using a value-based approach. They said: 'Those utilizing value metrics are growing at nearly double the rate as those deploying strictly feature differentiation.' They found those using a value-based approach were growing at double the rate of the others.

They then considered two types of value metrics:

- Functional—pricing scale around usage, e.g. price based on per user or per 100 videos
- Outcome-based—pricing based on results, e.g. price based on how many views a video received or how much money you made your customer

CEO, Patrick Campbell, said: 'When we look at the gross churn of companies using different models you'll note that both types of value metrics outperform feature differentiation with up to 75% less churn, but the outcome-based value metrics take this a step further with an additional 40% reduction in churn.'

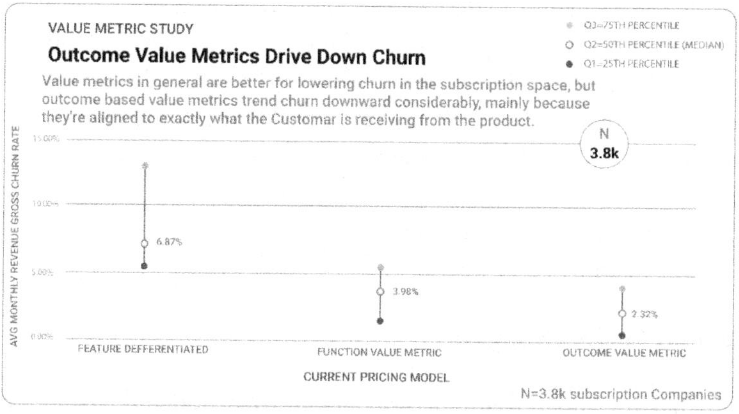

SOURCE: PROFITWELL VALUE METRIC IMPACT STUDY

Figure 84: An outcome focus reduces churn

He continued: 'This trend continues further when looking at expansion revenue. Both types of value metrics still outperform feature differentiated pricing models with at least 30% more expansion revenue, but outcome-based value metrics push those gains to nearly 50%.'

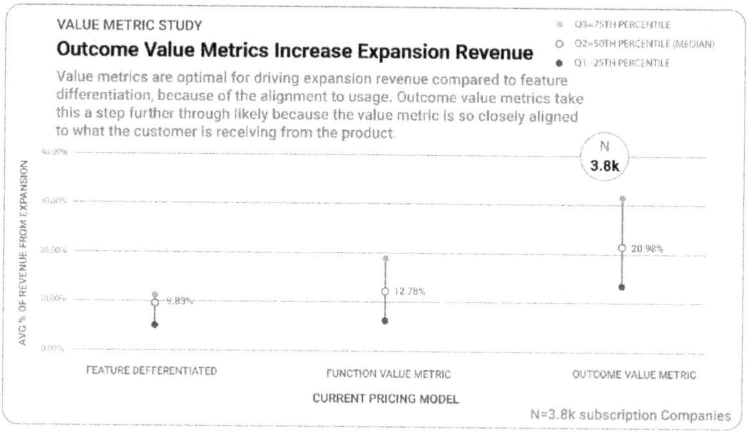

SOURCE: 2018 PROFITWELL VALUE METRIC IMPACT STUDY

Figure 85: An outcome focus increases expansion revenue

We made the point earlier that an outcome program doesn't necessitate the use of outcome-based pricing. But here's what Patrick Campbell concluded: 'We all don't have the luxury of pricing based on outcome though, because sometimes it's hard to perfectly measure how much money someone gained from using your product or how much that time you saved them is worth. Yet, we can take a lesson from this data in making sure we get as close to that customer and as close to value as technically possible.'

What This Analysis Suggests

The further you move along the pricing maturity model, the greater the financial benefit for a vendor.

The greatest benefit lies with the Pay-for-results (outcomes) model. But to benefit, a vendor must be capable at enabling business outcomes. If the customer doesn't achieve the outcomes, the vendor loses financially.

So, you need to build the capability to enable outcomes if you are to attain the strong financial benefits of Pay-for-results pricing.

Summary

Lead with the Success Outcome. It acts like a travel poster. And it invokes both WIIFM and WIIFC.

To increase price:

- Do more of what the customer needs to achieve the to-be state.
- Work jointly.
- Remove as much risk as possible.

The price a customer will pay depends on

- The value they perceive (WIIFM and WIIFC)
- The extent of your role
- The risk they perceive

The more you focus on enabling a to-be state Deciders see as high-value, the better you will fare.

And you can consider further moves along the path to outcome-based pricing.

PART 12 – Outcome-focused Offerings

What We'll Cover

An outcome program provides a great new lens through which to make decisions on your products and services. In Part 12, we will discuss how your products and services can change in support of your outcome program.

We'll cover:

- Outcomes as a frame of reference for your product roadmap
- Building your offering to enable your Success Outcome
- Bundles—turning outcomes into products

Section 1 – Outcomes as a Frame of Reference

Mature companies have a roadmap which sets out their development plans for the foreseeable future. Developing the roadmap can prove challenging. Each department wants input. Diverse points of view create differing ideas and priorities. And sometimes outright conflict.

The resulting roadmap can lack a clear direction. It has compromises so everyone feels they got something they wanted.

This approach sometimes creates tactical product roadmaps. And tactical roadmaps become fluid. Someone will offer a better idea than those currently under development. So, product development must change focus. Product development often has a string of unfinished projects.

A Common Frame of Reference

An outcome program offers a common frame of reference. It provides a single lens through which to consider priorities on product.

For a line of business, you enable a Success Outcome for customers. The Success Outcome is the North Star, the guiding light. The Success Outcome drives marketing messages, sales pitches, professional services methods and Customer Success (CS) engagement.

Product development should follow the same North Star.

Your outcome program should drive an increase in revenue and profit. You win the initial contract by showing the prospect the importance of the Success Outcome you offer. And you then show the prospect why you're the best company to enable that outcome for them.

You drive expansion revenue by identifying initiatives every year (or more often). The initiatives drive further improvements in the Success Outcome. The Success Outcome will never be 100% achieved. Opportunities for improvement will always exist.

So, finding new and better ways to enable the Success Outcome should drive product development. When a debate about competing priorities occurs, the simple question is, 'Which option best improves the Success

Outcome?' Of course, you need to balance the degree of difficulty and cost in providing the respective product developments. But all other things being equal, development should focus on the option that best enables or improves the Success Outcome.

Section 2 – Don't Build a Product, Build an Outcome

After the first release of a product, development is an incremental process. Each new release builds on the release that preceded it. And for the most part a new release delivers new features and functions (not always the case, sometimes a release may be technical in nature, e.g. a database change).

You have shifted your thinking around customer engagement from product to outcome. Your thinking around product development should follow the same course.

We discussed earlier that any business outcome requires PSPSPS elements. Product development should now consider all five of these elements when deciding how to improve the Success Outcome.

Processes

Perhaps your customers have antiquated processes. The customer may have developed their processes without conscious thought. Perhaps they followed the advice of the previous vendor that you've now replaced. And they have grown used to working with these old processes. Sometimes they may ask you to modify your product so that the old processes continue.

We need the customer to adopt more up-to-date processes to achieve the Success Outcome.

Your next improvement may not come from new features in the product. For example, you might create an education course on best process practices. You help the customer see their old and antiquated processes in a new light.

Or, you could partner with a company offering business process management tools. Have that company identify bottlenecks and process inefficiencies in the customer.

Or, you could create a simple self-analysis tool. Create a series of questions, the answers to which highlight where process ineffectiveness lies. Let the customers reach their own conclusions about the need to improve their processes.

Systems

Helping customers analyse their systems is your bread-and-butter. But even in this area, alternatives to new feature/function development may exist. For example, the customer's staff may use mobile phones more than management suspects. A simple survey to understand usage profiles may open the eyes of your customers.

People Skills

When we think about people skills, technology vendors tend to feel their job is to train customer staff in how to use the product. If the customer's staff know how to use the product, the vendor considers their job done.

But most vendors have experience with users who fail to adopt the product. Or who start, but drift away. The users stick with the old ways. They use manual processes that bypass the main system. Off-line spreadsheets appear. It's usually not malicious. Just easier.

People skills are not just about skill in using the product. Systems link people. People need to understand how they affect other people. They need to understand the broader picture of what the company wants to achieve.

People skills have a change management element. Introducing a new product creates business change. Even simple changes benefit from change management. The more complex the change, the more formal the change management needs to be.

The more complex your product, the more important that you work with the customer to drive change management. Poor change management reduces the likelihood of achieving or improving the Success Outcome. And if the customer doesn't feel successful, your revenue is at risk.

Investing in a change management program may prove better than a new round of features in the product.

Partners

If your customers have external partners playing a role in the outcome, the partners must be well integrated. Customers don't always get this right.

You can invest in guides, tools, self-analysis to help customers understand how to better integrate their external partners. Again, this may be a better investment than the next round of features and functions.

Supporting Elements

Supporting elements include policies, measurement of results, reporting arrangements and so on. You're helping your customers implement change in their business. Your customer will need to consider their policies. They may have policies that need to adapt.

Larger customers will have formal policies. In smaller organisations the policies may be far more informal. In both cases, it's easy for policies to become out of date and thus ignored.

Providing a guide on policies will increase the chance of success for your customers. Or you could offer them a template they can modify. Or an online tool to develop their own policy manual.

Customers might sideline measurement and reporting for future treatment. But the old adage of 'what gets measured, gets managed' is alive and well. If your customer does not have measurement and reporting that reinforces the new to-be state, the to-be state may slip away.

Ideally, you will build the measurement into your product. Let the product automatically capture the data that tells the Decider how they are performing.

You undoubtedly offer measurement and reporting for customers. But would greater investment have a more positive impact on your customer's success (and therefore your revenue)?

All these examples highlight a principle. You should develop an offering that best enables the Success Outcome you create for your customers.

So, you should not develop a product roadmap. You should develop an outcome roadmap.

Section 3 – Outcome Bundles

To achieve and then improve the Success Outcome, customers need to get PSPSPS right. As a vendor, you have an interest in helping with as much of this as possible.

Firstly, it increases the chance of success for the customer.

Second, it's an immediate revenue opportunity. You can expand your offering.

An Outcome Bundle consists of products, services, tools, guides, methodologies and partner offerings. Everything the customer needs to achieve the Success Outcome.

Developing Outcome Bundles

Think about everything a customer needs to achieve the Success Outcome—PSPSPS. Make a list of what the customer needs. Focus on the major elements only. You would expect to have 14–20 elements in total.

Figure 86: What the customer needs to achieve the Success Outcome

Now think about where the customer gets help on each element. For how many of the elements are you the primary source of help for the customer? Perhaps your list might look something like this, with a tick indicating elements you address:

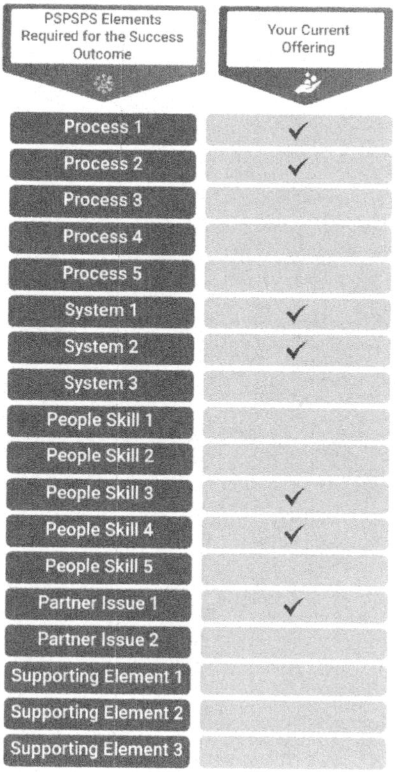

Figure 87: How much of what the customer needs you may currently offer

To drive income from outcomes, you need to help the customer with as much of PSPSPS as possible. You won't be able to do everything immediately. It will take time to develop your offering.

Your first Outcome Bundle might look like this:

PSPSPS Elements Required for the Success Outcome	Your Current Offering	Your First Outcome Bundle
Process 1	✓	✓
Process 2	✓	✓
Process 3		✓
Process 4		
Process 5		
System 1	✓	✓
System 2	✓	✓
System 3		
People Skill 1		✓
People Skill 2		
People Skill 3	✓	✓
People Skill 4	✓	✓
People Skill 5		
Partner Issue 1	✓	✓
Partner Issue 2		✓
Supporting Element 1		✓
Supporting Element 2		
Supporting Element 3		

Figure 88: Your first Outcome Bundle

Over time, you will increase the scope of the Outcome Bundle for this outcome. You may never achieve 100% of what the customer needs. Sometimes this will be your customer's choice. A future Outcome Bundle might look something like this:

PSPSPS Elements Required for the Success Outcome	Your Current Offering	Your First Outcome Bundle	A Future Outcome Bundle
Process 1	✓	✓	✓
Process 2	✓	✓	✓
Process 3		✓	✓
Process 4			✓
Process 5			✓
System 1	✓	✓	✓
System 2	✓	✓	✓
System 3			✓
People Skill 1		✓	✓
People Skill 2			✓
People Skill 3	✓	✓	✓
People Skill 4	✓	✓	✓
People Skill 5			
Partner Issue 1	✓	✓	✓
Partner Issue 2		✓	✓
Supporting Element 1		✓	✓
Supporting Element 2			
Supporting Element 3			

Figure 89: A future Outcome Bundle

Assembling an Outcome Bundle

You don't need to develop all components of an Outcome Bundle yourself. For example:

- If you need to expand how well you help customers with processes, you might find another software vendor who has an offering that fills a gap. You can partner with them.
- On people skills, you may have identified change management as a major gap affecting your customers' success. Again, you might find a partner. Or a methodology someone else has written you can license.

You want to assemble as much of what the customer needs as you can. And then build on it over time. And you'll have lots of options for getting there.

Outcome Bundles Help Your Sales Team Move to Outcomes

The sales team in most organisations has learned to sell products. Even those skilled in solution selling focus on how their products solve a set of customer problems. Now, you want the sales team to sell outcomes. For some, this will take some adjustment

Sales teams can use Business Maladies as a steppingstone from a problem approach to an outcome approach.

Outcome Bundles provide another method for easing sales teams into an outcome approach. You can develop a bundled offering that addresses what the customer needs to achieve an outcome.

You can then put a price on that bundle. An Outcome Bundle has all the characteristics of a product. It has features, a clearly defined offering, and it has a price.

And that should make your sales team more comfortable. Even though they will be talking about outcomes, they have a 'product' to offer.

Section 4 – Product-Enabled Outcomes

In Section 2, we suggested you should not build a product, you should build an outcome. You should aim to provide most of the PSPSPS elements needed to achieve a business outcome.

In Section 3, we discussed Outcome Bundles. In the short term, the additions you make to your Outcome Bundles won't come from traditional product development. For example, you may partner. Or build tools, guides or self-assessments. Or you may develop new professional services offerings.

That's OK for the short term. The long-term direction of product development should be to enable as much as possible of PSPSPS in the product.

The simpler the product, the more of PSPSPS the product must deliver.

In time, you create Product-Enabled Outcomes.

The Pendo platform creates software that self-diagnoses customer business outcomes. Ben Carey, Chief Customer Officer of Pendo, said:

Our customers write software. They want the usage of that software to grow.

We understand that companies pay for software to achieve a business outcome. We don't see completion of the software as the end of our role. We believe the software should play an active role in diagnosing and identifying ways to improve business outcomes. That's what the Pendo platform does.

We apply this thinking to our own customers. We have developed a product-led model. We are automating how we help our customers achieve and improve their business outcomes.

We've identified indicators that tell us whether our customers are likely to achieve their business outcome. When one of our customers isn't acting in a way that will lead to their success, we help them. For example, we'll automatically send them a short email, or provide in-app guidance, with a suggestion for improvement. The suggestion may not involve using the software. But our experience tells us it will increase their chance of success.

And we don't see achievement of their success as a one-time event. We see it is a continuous improvement journey. We continue analysing to further improve the business results our customers achieve. And we automate the way we help our customers as much as possible.

Our self-help model aims to help all customers, not just 'tech-touch' customers. We want all of our customers to achieve product-led outcomes.

Section 5 – Services Companies

Services offerings prove difficult to differentiate. In a customer's mind, one group of people offering a service looks like the next group of people offering that service. Whoever offers the lowest price wins the business.

Services companies can combat this by developing small offerings they add to the mix. Perhaps they are a services partner for a major application. They will have small applications that fill gaps in the major application.

Managed Services Providers, development companies and other providers of IT services may also have pre-developed software.

But a services company can also become known for enabling a Success Outcome. And they can become so good at enabling this outcome, that they stop competing as if they were offering commodity services. They begin to compete on their ability to deliver measurable business results for their customers.

For many services providers, committing to a single business outcome might seem frightening. They often trade on their ability to do whatever the customer wants. But in the end, regardless of how broad the range of services offered, the customers still consider them a commodity.

Back to the medical analogy. A heart surgeon makes far more money than a general practitioner. And if you need heart surgery, you want to engage the best heart surgeon you can find. For as long as people have heart conditions, the heart specialists will have ample work. At high margins.

A services company that chooses a Success Outcome can develop their offering. They can include everything the customer requires to achieve that Success Outcome. They can create Outcome Bundles in the same way a product company can.

PART 13 – Outcomes and Reducing Costs

So far, we have mostly concentrated on the revenue side of *Income from Outcomes*. But an outcome program can also reduce costs.

Section 1 – Sales Costs in Existing Customers

Face-to-face (Larger) Existing Customers

Traditional face-to-face selling proves expensive for most technology vendors. The major steps look something like this:

- Qualify—the account manager will ask questions such as:
  - Do they have budget?
  - Is there a compelling reason for change, and does the change need to happen now?
  - Can we solve the problems they have—are we a good fit?
- Discovery—understand the problems they face.
- Demonstration—present and then demonstrate how your product solves the problems.
- Negotiate and close—agree commercial terms and get contracts signed.

An account manager plays a role from start to finish. Good account managers are expensive. And there's typically a strong emphasis on demonstration of software to prove you have a solution to the customer's problems.

To do demonstrations well, you need great pre-sales people. And they not only have to do the demonstrations, they must also spend significant time in setup. Demonstrations are time-consuming and resource intensive for pre-sales people. And great pre-sales people are also expensive.

The cost of all this face-to-face engagement is borne by the vendor. It's an expensive process.

Now consider the IDEEAL Lifecycle model. The major steps are:

- Through an Executive Business Review (EBR), identify possible areas for business improvement.
- Jointly develop a high-level path to success.
- Jointly conduct a deep study on a business initiative. Then write a report with everything needed to get the business initiative (a project) approved.
- Have the customer select you for the project.
- Negotiate and close.

The reasons this model proves less expensive than the traditional selling model include:

Paid Pre-sales

The most time-consuming activity in the BuILD process is the deep study. But customers should pay for your work on this study. This is the 10% Sale. And because you develop a product agnostic report, the consulting rates you charge can and should be high.

So, the most time-consuming element of what would be considered a pre-sales phase is a paid engagement at high margins.

Few Or No Demonstrations

With existing customers, it's likely you won't need demonstrations. You will build faith through your BuILD process.

Customers will ask if your products support the project scope. And they will believe you when you tell them your products will do what they need.

Some customers may ask you for a proof-of-concept demonstration. But this won't be a 'sales exercise'. You'll be confirming what they already believe—your products will do what they need.

Avoiding 'Do-nothing' Decisions

'Do-nothing' decisions are the bane of most sales teams. The sales team goes through a lengthy sales process with the customer, only for the customer to decide not to go ahead with anyone. They decide to do nothing. The sales cycle has proved a waste of time. Often, sales teams lose more deals to 'Do-nothing' decisions than to any other competitor.

The 10% Sale solves that problem. Early in the engagement, you're asking the customer to pay to conduct a deep study. The study provides strong confidence and reduces risk for the customer.

If a customer will not pay 10% to achieve that level of certainty, it's unlikely they were ever serious about the business initiative. It's much better to find that out early in the process rather than after a lengthy sales cycle.

The 10% Sale substantially reduces the number of occasions you will suffer a 'do-nothing' decision. And that will lift the productivity of your sales team, reducing average cost per sale.

Reduced Time-to-decision

An interesting thing can occur as customers get used to your BuILD process, particularly the 10% Sale. Customers see the value of conducting a deep study.

Now, whenever a customer raises a problem, account managers will propose a paid study (the 10% Sale).

For those that go ahead with the study, they get a report with everything needed for approval. And that reduces hesitancy. The customer has confidence to submit a proposal for the business initiative. And it's likely to receive approval.

That process sharply reduces the average sales cycle.

And customers who have enjoyed success using a study will begin to proactively engage you. They will call you. And ask to pay you to do a deep study. You will work jointly with the customer on the outcome report, submit that report and on most occasions the initiative will go ahead. It moves quickly.

The cost of sale is low as the customer pays for the bulk of the engagement. And it moves faster than a traditional sales cycle.

Self-Help Customers

Self-help customers often present vendors with a conundrum. There's enough revenue in total from these customers to make them valuable. But the revenue from individual customers is too small to support the cost of face-to-face engagement.

For many vendors, it becomes too hard. And so, they tend to ignore these customers unless the customers contact them.

A program to automate your BuILD process has a great chance of creating revenue from self-help customers. You'll need to make an initial investment to set up the automated systems. But once done, you will have an engine to drive leads from this set of customers. At very low ongoing cost.

Section 2 – Sales Costs in New Business (New Logo) Deals

Win-Rate

We have discussed how to use your outcome program to differentiate. Your Success Outcome will be unique. You will compete on your ability to help the customer with everything needed to achieve the Success Outcome. This new foundation for competing and winning deals should improve your win-rate.

And improving your win-rate will decrease your average CAC (Customer Acquisition Cost).

The Cost of Building Faith

The traditional selling model has an unspoken assumption that building faith will fall mainly to the sales team. But prospects now engage vendors late in their buying cycles. The opportunity for sales to build faith has plummeted.

Your BuILD process offers a new way to build faith. For New Business prospects, you will offer much of your BuILD process as self-help. You don't yet have the relationship to run this as a face-to-face model.

But if you run your BuILD process well with New Business prospects, you will build faith with them. By the time they are ready to speak to your sales team, the prospect should feel they know you and can trust you already. The sales team can then build on that faith, rather than starting from zero.

And that should mean the amount of 'relationship-building' time and cost will reduce. And because prospects already have some faith in you, you'll spend less time 'proving' everything you say. Again, this will reduce cost.

The 10% Sale

So far, we have treated the 10% Sale as a step you run with existing customers. But the underlying concept of completing a deep study addressing all elements of PSPSPS can apply to New Business prospects.

It's possible to convince New Business (New Logo) prospects to engage you through a 10% Sale.

When you're successful, the 10% Sale replaces costly vendor-funded pre-sales with a paid engagement. And that brings the customer acquisition cost down.

Section 3 – Services Costs

When a project fails, customers want to blame someone. And vendors provide a soft target. If the customer saw the project as an IT-style project led by the vendor, then the vendor will be held accountable.

The customer will demand the vendor take corrective action at their own expense. And that means providing services free of charge.

An outcome approach changes things:

- You agree a measurable objective for a project. From the start, everyone has clear expectations about what the project will deliver. This reduces the chance of the customer having the wrong expectations and blaming the vendor.
- Because you're working on all aspects of PSPSPS, it must be a joint project. It's not a 'vendor' project. If something goes wrong, it's a shared responsibility. This reduces the demands that the vendor fix any issues alone.

Section 4 – Paid Customer Success/Customer Support

It's not uncommon for customer support departments to offer multiple levels of service. The basic service may be free or included in the cost of maintenance or subscription. Higher levels of service are available at a cost to the customer. The service levels typically vary around time-to-respond and dedicated resources.

We see Customer Success (CS) departments also offering different paid services.

Outcomes offer a new opportunity to create paid services. For example, SaaS vendors have unprecedented access to their customers' data. With each customer's agreement, they can monitor key outcomes and provide meaningful reports on a regular basis. It may be possible to set up alerts when an outcome appears to be under threat. And customers could pay for these services.

Paid services can offset the costs of operating a customer support or CS department.

Section 5 – Personnel Costs

An outcome program creates a stronger sense of purpose for staff. Having a clear link between the work the staff do and measurable business outcomes for customers creates this higher purpose.

A higher sense of purpose has three positive impacts for a vendor:

- Staff retention improves—people want to stay with an employer when they have that strong sense of purpose.
- Discretionary effort—people who have that stronger sense of purpose will work harder. This means the output from a team is higher, or the team can be smaller to deliver the same output.
- Recruitment—having the strong sense of purpose will attract people seeking a strong purpose. Staff will comment positively about this purpose in online feedback sites. And this will make it easier to recruit high-quality staff.

All three benefits can reduce personnel costs.

PART 14 – How to Proceed With an Outcome Program

At some point, all customer-facing departments should adopt your outcome approach. And you'll need R&D/Product Development to follow suit.

But you don't need everyone to take part from the start. You should see your outcome program as an evolution, not a revolution.

It's likely you're already doing some of the things addressed by an outcome program. You may not have thought of those things as outcome-focused. But they will be there. You'll have a starting point on which to build.

You'll go through a design process. You'll come up with your approach and start small. You'll then learn and adapt your program. And progressively expand it.

In Part 14, we will provide suggestions about how to proceed with your outcome program.

Section 1 – Design Your Outcome Program

Keep the Six Decisions in Mind

In designing your program, you should focus on the six decisions you need customers to make. And the emotions needed for those decisions to take place.

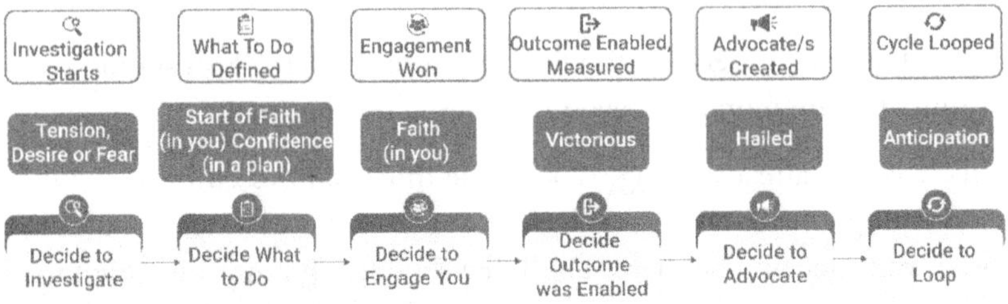

Figure 90: The six steps, emotions and decisions

Consider at each step whether your actions will create strong emotions. Strong enough to reach the Action Threshold.

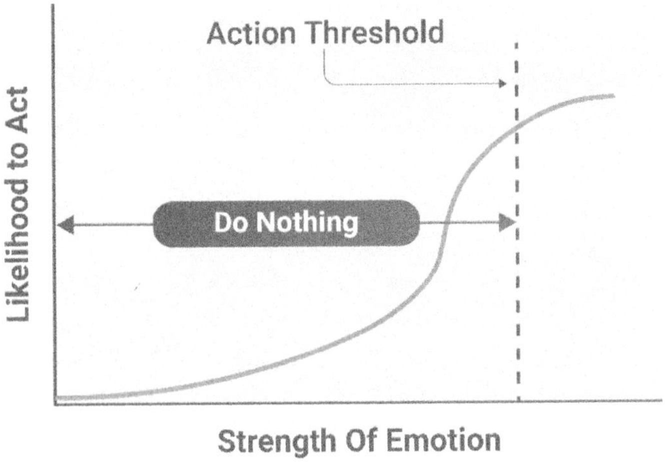

Figure 91: Emotions must reach the Action Threshold

Choose Your Starting Point

While you might define your outcome program for all segments, it's more likely you'll choose one segment to pilot. You may, for example, choose your face-to-face segment (larger customers).

You need only one Success Outcome for a line of business. If you have more than one line of business, choose the line of business you'll start with.

You don't need to involve all staff in the initial rollout. You'll find your outcome program resonates with some staff immediately. Other staff may need a little more convincing. That's OK. If possible, begin with the enthusiasts.

When it's time to begin the rollout, you won't involve all customers in your chosen segment right away. Pick some early adopters to work with.

And similarly with partners. If partners will form part of your program, pick those most likely to succeed.

With each of these choices, choose the approach that offers the greatest likelihood of success. Early success will create momentum for your program.

Define Your Outcome Structure

Your Success Outcome is your guiding light, your North Star. It describes what success looks like for your customers.

For a line of business, you only need one Success Outcome. It makes sense to work that out first. It will influence the rest of your design process. When you have defined your Success Outcome, you then establish your Contributing Outcomes.

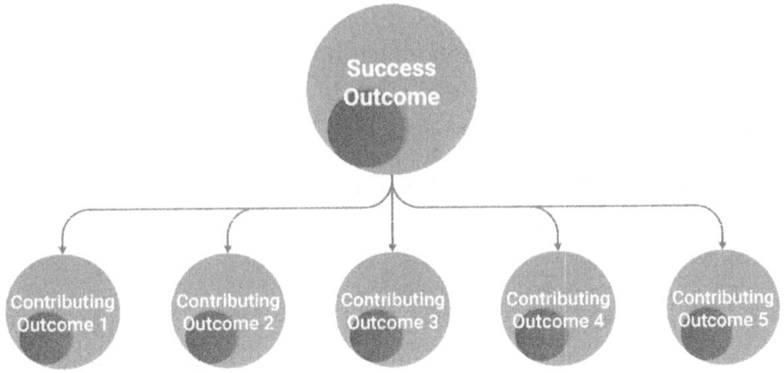

Figure 92: Your Outcome Structure is your centrepiece

Your Outcome Structure provides the centrepiece for everything you do with your outcome program.

Define Who You Engage

You need a clear idea of who you'll engage in your chosen segment.

Decider

Put most of your focus on the Decider. Until a Decider Decides to Investigate, there's no buying cycle. The Decider will shepherd the budding business initiative through each phase.

You may have many Deciders. But there's usually a Lead Decider—the person who takes the initiative and Decides to Investigate.

Approver(s)

You need to identify the usual approving body. It may be a board, an executive committee or a single executive with enough authority.

Other customer people may need to approve a business initiative. These Approvers won't have the authority to approve the initiative, but they will have the authority to veto it. The obvious starting point are the people above the Decider in the Outcome Hierarchy.

Approvers may also exist outside the direct 'chain of command' above your Decider. For example, initiatives with technology elements may need sign-off by the IT department.

When designing your program, you not only need to activate your Decider. You need to think through how you'll activate the Approver(s).

Recommenders

In a typical engagement with customers in your chosen segment, who else will play an influencing role? Who might the Decider consult? Or who would the Decider need to get on board, even if they don't have a formal approval role?

Executors

Once a business initiative obtains approval, who will the key players in the project be? How will you engage with them during the BuILD and Deal Won phases?

Once you're clear on who you will engage in this segment, you can design how you'll engage with them.

Pathway 1 – Start with Your BuILD Process in Face-to-face Customers

Your largest and probably most important customers will be a natural focus for your outcome program. Begin by designing how you'll execute your BuILD process in this segment.

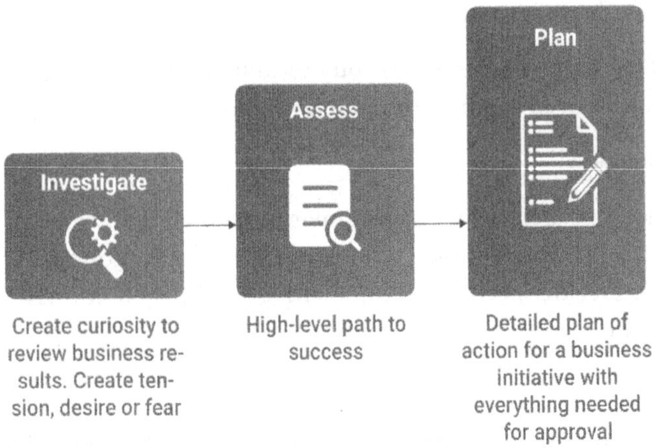

Figure 93: Your BuILD process

As you design your program, think through how you'll create the necessary emotions. You need the Decider to feel:

- Tension, desire or fear to Decide to Investigate, then
- Some faith in you (to listen to you) and confidence to Decide What to Do, then
- Faith in you when they Decide to Engage You.

With a design in place, start a trial. Talk with a few select customers and gauge their interest in jointly engaging on the Success Outcome. Some customers will immediately love it. Others will feel hesitant. And still others may decline (at least for now). Choose the most positive customers.

Then, agree how you will engage. If you'll use EBRs, agree the purpose, attendees, format and timing with each participating customer.

And then get started.

In parallel, start work on designing your outcome-based project management approach. When customers sign up for an outcome-driven business initiative, they will expect you to guide them. Be ready.

When you achieve success, implement your advocate program by turning the key customer people into heroes and heroines.

Finally, get the flywheel of momentum turning. Agree the process and timing for looping the cycle you've just been through with this customer.

Pathway 2 – Start with your BuILD Process in Self-Help Customers

You might decide to start here because engaging self-help customers presents the biggest gap in your customer engagement. Or because the bulk (or all) of your customers are self-help.

The BuILD process has the same steps in self-help customers, but executed electronically.

You should consider how you'll create tension, desire or fear in the Decider in these customers. A report setting out their performance, possibly with benchmark data, provides a possibility. But you may think of others. For example, a periodic webinar where you discuss industry trends and benchmarking data for the Deciders.

You'll then need to help the Decider self-assess where the opportunities for improvement might lie. A questionnaire could do this.

The Decider will then need a detailed plan. Ideally, you'll provide tools which will allow self-help customers to develop the plan themselves.

If your product requires implementation services, you'll lead an outcome-based project. If your product doesn't require implementation services, you'll want the product to deliver as much of the outcome as possible.

You'll still want to create heroes and heroines. And you'll then want to automate the looping of the cycle.

Pathway 3 – Begin With Outcome-based Projects

You might start here to build the expertise to deliver first. Or because you have deals with existing customers you could convert to an outcome approach.

Eventually, you will need references and hopefully advocates for your outcome program. You might decide to start by creating them.

You now have a plan for who you will engage and how you will engage. Next, you can plan how to launch your outcome program.

Section 2 – Plan Your Launch

Scope and Objectives

Eventually, all departments need to take part in your outcome program. But you need to start somewhere. If you start with the BuILD process, who will lead it? Perhaps sales, or sales plus marketing, or Customer Success (if you have a CS team).

Make your choice based on likelihood of success.

You may also choose to set some objectives for your outcome program. How many customers you will engage, how many business initiatives you want to create, perhaps how much revenue you'd like to close?

Success Consulting

The success of your outcome program increases through Success Consulting. You help the customer with all aspects they need to achieve the Success Outcome. You need people who understand all aspects of PSPSPS.

You need to decide who will carry out the Success Consulting.

Execution Capability

For your outcome program to succeed, you need execution capability, i.e. the ability to execute the plan that you develop. For customers to achieve the Success Outcome, they need the right PSPSPS.

For your outcome program to succeed, you need to develop PSPSPS internally.

You should consider the processes, systems, people skills, partners and supporting elements needed to execute each step of the Outcome Lifecycle. And then develop a plan for how you will create this execution capability internally.

Change Management

The likelihood of success will increase if you adopt a change management approach. This is an outline of what you should address[5]:

Topic	Questions
Need for Change	Why is the change necessary or desirable?
Vision	What is the 'big idea' behind the change? What do we want to achieve? What is the To-Be State?
Strategy	How will the vision be achieved? What are the highest priorities?
Resources	How much will the change cost? Who will be working on the program? For how long? Will there be offsetting benefits?
Executive buy-in	Commitment from which managers/executives is required? What is the current status?
Scope	What functions and/or activities are included in this change? Are there areas specifically excluded?
Time frame	When will this change start and when will it be completed?
What will change?	Who will be affected by the change and how? Is there a change to the organisation structure? What changes to processes, systems, people skills or partners required? Transition arrangments needed?
Risks and restraining forces	What could adversely affect success? For significant risks and restraints, what contingencies are there?
Consolidate wins/Evaluate progress	When, how and by whom will regular evaluations during the change be made?
Anchor change	How will success of the program be measured? When and who will do this?

Figure 94: Change Management outline

Choose the Participants

Choose the staff you'll involve at the start.

Choose the customers you'll approach to enrol in your program.

Choose any partners you need to take part.

Launch

You're now ready. You can launch your program.

CONCLUSION

You now have a framework on which to design and run an outcome program. Or to consider next steps if you already have an outcome program.

Let's review:

Section 1 – Summary of Key Concepts – Book 1

What You Provide

For a line of business, you enable a single Success Outcome and four to eight Contributing Outcomes for customers. That's it. This simple structure underpins everything you do for customers.

To achieve an outcome, customers need the right PSPSPS. Your future revenue depends on customers getting all five of these elements right.

Who You Engage

You should consider the Outcome Hierarchy of your customers. Each role in the Outcome Hierarchy has a key outcome it enables. You will find the role that most closely aligns with your Success Outcome. This is the person most likely to Decide to Investigate a business initiative that will need your products and services.

This role is your Decider. If you can identify a role that starts a business initiative 80% of the time, you have enough to build your program. The Decider will make most or all of the six IDEEAL Decisions.

How you Engage and Deliver

Customers must make six decisions for you and the customer to prosper together. If they fail to make any of the decisions, their journey stops. At least with you.

To make those decisions, customers must feel the right emotion at each step. Six decisions, six emotions. And the emotion must be strong enough to reach the Action Threshold.

So, you need to design the six steps of your IDEEAL Lifecycle with just one purpose—to create the six emotions to elicit the six IDEEAL Decisions.

Six steps to create six emotions to elicit six decisions.

Section 2 – Summary of Book 2 - Making the Concepts Work

Lead Generation

Your outcome program will overcome the void in customer engagement. You'll build a pipeline of business initiatives. And use the BuILD process to create this pipeline.

This will underpin your expansion revenue program.

You can also use BuILD to create a pipeline of New Business (New Logo) prospects.

Sales

If existing customers have been through your BuILD process, you're unlikely to have a competitive sales situation. You have created faith through your phase-appropriate engagement. The deal will be yours to lose.

In New Business, you'll differentiate and reach the right executives using your Success Outcome. If a prospect contacts you late in their buying cycle, you can disrupt their buying process by bringing them back into your BuILD process.

Outcome Enabled, Measured

You've sold a business initiative. Now you and the customer must deliver. You'll use an outcome-based project management approach. You will focus on delivering a measurable business result from each project—a project outcome.

Advocates Created

When the business initiative succeeds, you create heroes and heroines in the customer. They will become your advocates.

Cycle Loops

Continuous improvement of the Success Outcome will underpin your revenue expansion program. You'll loop back and go through the cycle again and again.

Outcome Techniques

Your outcome program will utilise techniques that will help set you apart:

- Success Consulting
- The 10% Sale
- Your Business Malady
- Dealing with multiple Deciders in customers and prospects

Outcome-focused Offerings

You'll build more of what the customer needs to achieve success (PSPSPS) through Outcome Bundles. And over time, you'll deliver more and more of your Success Outcome through your products.

Evolution Not Revolution

When beginning your outcome program, start small, learn, build success and then grow over time.

Section 3 – Unrealised Income

An outcome program should help you tap into Unrealised Income. The extra revenue and profit that's out there.

These are the eight problems that cause Unrealised Income and how Income from Outcomes has addressed them.

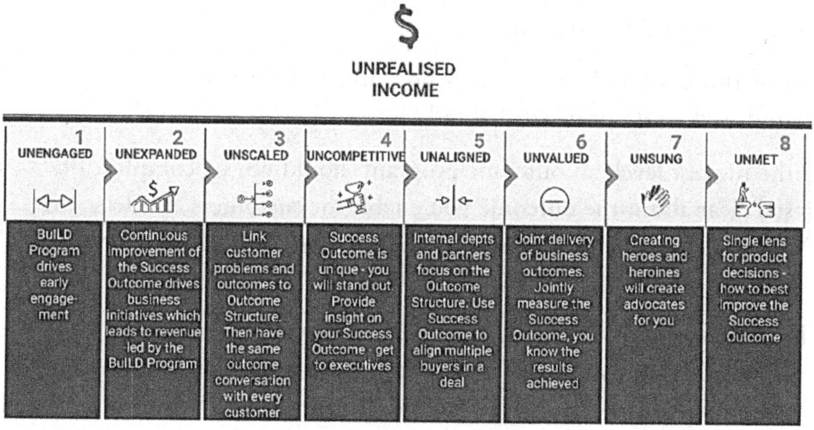

Figure 95: How Income from Outcomes addresses the eight Un-Problems

END WORDS

Over the past few years, I've seen more and more written about the need to deliver business outcomes for customers. And I see more and more companies launching programs with this in mind. The groundswell for this approach gains momentum.

I wrote this book to help anyone who wants to be part of that groundswell, and not be left behind.

At the highest level, an outcome program should not be complex. Be crystal clear about the outcome you enable for customers. Understand what the customers need to achieve that outcome. Then, do everything possible to ensure they achieve the outcome.

The challenge with this book was the amount of detail to include. My first draft had enough words for two books. I wanted to include enough detail to assist someone designing their own outcome program. But not so much that the book would be overwhelming. Hopefully, you find the balance acceptable.

I wish you great success in your own outcome journey. I hope the book proves of value to you along the way.

NEXT STEPS

Website

You'll find more content on customer outcomes at https://outcomeleaders.com/

Help

At Outcome Leaders, we provide training, workshops and consulting to help you set up your outcome program. Or to help take your current program to the next level. You'll find more information on the website.

Contact Us

We'd love to speak with you. The simplest path for you is to use the Contact Us form on the website.

APPENDICES

Appendix 1 – Setting Up Your Personae

To assist with planning your outcome program, you'll find it useful to set up personae, or identities, for each of the key roles you engage.

The following approach will assist you.

'Real' People

Having a persona for your Decider will provide the most help. But you may choose to create personae for the other roles.

Turn each persona into someone 'real' by giving each persona a name—Julie, Patricia, Kale, Raji, Min, Gustav, Guillermo, whatever you like. Then refer to them by that name and not their job title. Find a photo of someone you think represents each of the personae and include it in any internal presentation.

What You Need to Know

You can now get to know each of the key people/personae. Consider:

Personal Attributes

How old are they?

What education do they have?

Do they have a family? If so, how many children?

What do they do on weekends?

What interests or hobbies do they have?

How do they feel about change?

Their Company

What is the size of the company—revenue and employees?

How does the company make money?

Who are their customers?

In which geographies does the company operate?

Their Job

What is their title?

Who do they report to?

How are they measured?

How many people report to them?

What roles do the people reporting to them hold?

What technology do they use?

What do they like about their job?

What don't they like about their job?

Career

How long have they been with their current employer?

How long have they been in the current job?

What are their career aspirations?

Sources of Information

General news

Job-specific information

Industry information

Consider formal and informal sources of information

Influencers

Who influences their thinking?

What's the attitude of these influencers towards what you do?

Please don't feel limited by this list—you may identify other characteristics that help you understand the key people with whom you engage. If so, add the new characteristics to the list.

Appendix 2 – Measurable Project Objectives

Transitioning to outcome-based project management can present challenges. These are some of the challenges you might face, and suggestions on how to respond.

Reticence from the Project Teams

In the past, project teams (vendor and customer) celebrated success on go-live. Now, you're asking the teams to achieve a successful go-live, and to ensure measurable business improvements as well. It adds risk and pressure for the project teams. You can expect some reticence on the part of the project teams.

Responding to Your Project Team

Your project team will have attended staff meetings at which you explained the reasons for adopting an outcome approach. They'll be aware of the change. Your project team should understand the benefits to the company.

You now need to help them understand why it's in their own interest to adopt an outcome approach to projects.

The first reason revolves around a higher purpose. Achieving a successful go-live provides satisfaction for the project team. It's likely the customer will enjoy some benefits. But the real benefits for the customer aren't measured. The project team doesn't really know if they helped the customer.

By setting measurable business objectives, the benefit of the project to the customer has clarity. Measuring results at the end ensures the project team knows exactly how much they have helped the customer. The link between their own hard work and measured benefits for the customer becomes clearer. And that should create a greater sense of purpose.

The second reason the project team should embrace the outcome approach lies with their own skill and experience. It's one thing to get a product working. It's a much bigger thing to deliver a measurable business result. This increases the skill level of the project team.

Third, the outcome approach opens new career opportunities for the project team. Internal promotion opportunities will arise. For example, Application Consultants may become Success Consultants. And, a track record of delivering measurable business results from projects makes them more attractive in the job market.

Of course, you won't want them to leave. But from a career perspective embracing an outcome approach will enhance their opportunities.

Too Big a Step

Your project team may feel overwhelmed. It all seems too much, too fast.

You should assure the project team you're adopting an evolutionary, not revolutionary, approach. You know they can't change overnight. So, you set out a plan to evolve, over time, to a full outcome approach. For example, you can first focus on setting objectives. Change the team's measurement and bonus schemes to reward them for establishing project objectives. Later, you can include measurement at the end of the project as part of their bonus scheme. At some point in the future, you may reward them on whether they achieved the project objectives.

Slipping Into Old Habits

Under the usual project pressures, it's easy for both your team and the customer team to slip back into old habits. They lose sight of achieving the business objectives under the day-to-day pressure of the project.

The leadership of your professional services team plays a key role in advancing your outcome program. Most services leaders review projects regularly. In the past, the focus has been whether go-live will occur on time and on budget. Now, the additional focus of achieving the business result must form part of the regular reviews.

And services leaders must ensure corrective action is taken if the project outcomes are at risk … even if the project is on track to be on time and on budget for go-live.

Appendix 3 – Success Likelihood Score Calculation

Scoring Your Customers

In Part 9, Section 3 of Book 2, we introduced the Success Likelihood Score. You can use this tool to assess customers and prospects. It will help you understand the likelihood an individual customer will enjoy success with you.

A simple spreadsheet will make it easy to complete your success likelihood review. Create a score for each of the three perspectives:

- Business Fit
- Project Success
- Expansion Potential

You'll find an explanation of the topics under each perspective a little later in this appendix. First, let's look at the Long-term View. Here's an example of a completed score sheet:

Perspective	Topic	Importance (A)	Score (B)	Weighted Score (AxB)	Maximum Possible Score (Ax10)	Comments
Business Fit	Alignment	4	8	32	40	The core outcome we enable aligns well with their strategic direction
	Viable To-be State	4	8	32	40	Clear understanding of how this part of the business should run
	Executive Engagement	4	7	28	40	Most key executives engaged
	Customer Financial Position	3	8	8	10	Strong financial position
	Business Relationship	2	6	12	20	Slightly adversarial
	Sub-Total			112	150	Score of 112/150= **75%**

Figure 96: Success Likelihood Score – business fit

Here's how the scoring works:

- Column A—a score for importance of each topic. For example, in Business Fit, how important is executive engagement? You will set the importance score once, and then use it for all customers. We suggest using a range of 1 to 4, with 1 for low importance and 4 for high importance for success. You can use a different range if you wish.
- Column B—a score for a customer on each topic. We suggest using a range of 1 to 10 to rate the customer on a topic. It allows for a more granular score.
- Column C—a calculation of weighted score. This will multiply the importance of a topic (Column A) by the score you give an individual customer (Column B).
- Column D—a calculation of maximum possible score—if a customer had a perfect score for this topic, what would that score be? Multiply Column A (importance) by 10 (the maximum score a customer can achieve).
- Comments Column—a place to make notes on why you chose the score for the customer on this topic.

The overall score is the total customer score (the sum of scores in Column C) divided by the maximum score (the sum of Column D).

The higher the resulting percentage score, the more likely you have a good business fit with this customer.

Once you have a spreadsheet template set up, you need only enter scores in Column B, i.e. score for this customer on a topic. And add some comments. The importance scores are fixed, and the other columns are automatically calculated.

It's a simple exercise.

Here's an example of the score sheet for Project Success:

Perspective	Topic	Importance (A)	Score (B)	Weighted Score (AxB)	Maximum Possible Score (Ax10)	Comments
Project Success	Project Objective	3	8	24	30	Measurable targets in place
	Funding	4	7	28	40	Budget a little tight, but OK
	All Elements Included (PSPSPS)	3	5	15	30	Inadequade focus on people skills, all other areas OK
	Governance	3	6	18	30	Authority levels not defined
	Project Management	4	7	28	40	Some gaps in project plan
	Project Staffing	3	8	24	30	Good customer team, with time available
	Stability	1	9	9	10	No significant risks
	Change Management	3	7	21	30	Some executives not convinced on need for change management
	Product Adoption	2	5	10	20	Ongoing review process not agreed
	Vendor Preparedness	2	8	16	20	Good fit for this project with right staffing available
	Sub-total			193	280	Acore of 193/280= **69%**

Figure 97: Success Likelihood Score – project success

You can see it follows the same structure.

And here's an example for Expansion Potential:

Perspective	Topic	Importance (A)	Score (B)	Weighted Score (AxB)	Maximum Possible Score (Ax10)	Comments
Expansion Potential	Measurement	3	8	24	30	Agreed process for ongoing Measurement
	People Skills Maintained	4	7	28	40	Plan for regular review
	Continuous Improvement	3	7	21	30	Have an informal program in place
	Change Adaptation	3	5	15	30	Seem slow to make changes and adapt
	Sub total			88	130	Score of 88/130= **68%**

Figure 98: Success Likelihood Score – expansion potential

For any of these templates, you may identify other topics more relevant to the Success Likelihood Score. You can change these templates as you wish.

Determining the Right Success Likelihood Scores

Unfortunately, there's no simple rule on what makes an acceptable score. It's a subjective judgement affected by factors such as:

- Likelihood of churn—if it's easy for a customer to switch to another vendor, you'll need a higher Success Likelihood Score.
- Potential financial loss—if the customer acquisition cost is higher than the revenue in your first contract period, you'll need a higher Success Likelihood Score.
- Lifetime value—if customers stay with you for long periods of time and your lifetime value is high, you might accept a lower Success Likelihood Score. The long-term revenue will be high if they don't churn.
- Your current business performance—in an ideal world, you would make the same decisions about proceeding with opportunities in both good and bad times. But with high quarter or financial year-end pressure, you may choose to take an account with a lower Success Likelihood Score.

You can use a retrospective approach to establish the right scores for your business. Go back over your customers. Attempt to score each customer as if you were just beginning the engagement with them, i.e. with what you knew about them at that time.

Are there any patterns that emerge? Are there threshold scores which give a good indication of a lack of success? Are there any factors which appear to be more influential than others?

Zones

You may choose to establish zones of success likelihood. You can use a traffic light approach. For example:

- Green means the likelihood of success is high—you will always engage with these accounts.

- Red means the likelihood of success is low—you have a policy decision to make on how you will treat your red opportunities. The simplest approach may involve an escalation process. Staff should escalate the decision on whether to proceed to higher management.
- Amber suggests there is moderate risk—again, you will need a policy for how you deal with amber accounts.

Once you have established your initial threshold scores and/or zones, you can monitor how well they predict success and failure over time. And adjust the thresholds as needed.

An Explanation of Each Topic

For each of the three perspectives on success, here's an explanation of each topic.

Business Fit

Alignment—is our Success Outcome a good fit for their business needs? Does it align with their strategic direction?

Viable to-be state—our Success Outcome describes a to-be state we will help customers achieve. Is the customer clear about this to-be state? Do they have realistic expectations? And do they understand what it will take to achieve the to-be state?

Executive Engagement—do we have engagement with people who:

- Have authority to make decisions on policy?
- Can authorise expenditure?
- Provide strategic guidance?

Financial Position of Customer—does the customer have ongoing funding, long-term profitability and positive cash flow?

Business relationship—is the likely relationship adversarial or collaborative?

Project Success

Clarity of objective of this project—does the project have measurable objectives agreed by all parties? Has the customer allocated enough funding to achieve these objectives?

All business elements included—the degree to which the customer has considered all elements for success. Does the project consider PSPSPS—processes, systems, people skills, partners, supporting elements?

Governance:

- Processes—are clear processes for review by a governance body in place?
- Authority—does the governance body have authority to make necessary decisions during the project?
- Executive engagement—are the right executives involved in the project?
- Policies—are they clear on policy changes needed to support the business changes this project will drive?

Project management—factors to consider include:

- Project manager quality
    - Customer and vendor
- Plan quality, including
    - Clear assignment of tasks
    - Realistic timing
- Plan focused on business results
    - As opposed to software go-live
- Commitment to project deadlines
    - Customer and vendor
- Consideration of partners and third-party involvement
    - Customer business partners
    - Vendor business partners
    - Other third parties

- Measurement of results achieved
    - Timing
    - Responsibility
- Project Staffing
    - Customer
    - Time available
    - Experience
    - Commitment/motivation
- Stability against adverse change
    - Customer business change
    - Customer marketplace change
    - Regulatory change
    - Project change
- Staffing
    - Funding
- Change management
    - Size of the change in the business
    - Degree to which the size of the change is understood
    - Begin with the end in mind
    - Formal change management process to be used
    - Clarity of change management plan
        - Initially
        - Over time
- Product adoption
    - Process for reviewing and improving adoption
    - Use of all licences/subscriptions
    - Advanced features
- Vendor Preparedness
    - Experience in running a project like this
    - Product
    - Closeness of fit of software
    - Degree of modification envisaged

    - Stability of required software version
- Staff
    - Adequacy of time funded by the customer
    - Individuals' experience in similar projects
    - Availability of skilled resources over the project timeframe

Expansion Potential

Measurement—is ongoing measurement in place for the Success Outcome and Contributing Outcomes?

People skills—does the customer maintain the necessary skill levels over time? How do they handle staff changes and onboard new staff? Do they provide ongoing training?

Continuous Improvement—do they have a desire for, and practice of, improving business results over time?

Change Adaptation—how well do they manage ongoing change internally and deal with external change? Are they interested in new ways to improve the Success Outcome?

ABOUT THE AUTHOR

Before founding Outcome Leaders, Paul ran the Asia Pacific region for a medium-sized ERP software company. He had 200 people across nine countries supporting 800 enterprise customers.

In the last five years he spent in that role, two interesting things happened. First, the company globally moved from on-premise to cloud or SaaS. He experienced first-hand what it takes to make a change like this.

Second, Paul decided that the Asia Pacific business should do more to create great business outcomes for customers. He saw this as a way to compete with the '800lb gorillas' of SAP, Oracle, etc. and to grow revenue.

So, he hired some smart people to complement the smart people already in the company. Together, they designed and successfully ran an outcome program for over five years.

At the end of that time, he realised a few things. First, customer outcomes had hooked him. He found the topic captivating and wanted to spend more time on the subject. Second, he started seeing more technology companies interested in customer outcomes. And thought he could help.

Finally, he realised he cared about the topic. He holds the view that it is not OK for companies to sell their products and services and for customers NOT to get great business outcomes.

He was working on his first book, *The Chief Capability Officer*, as something of a hobby. He resigned from the ERP company, set up Outcome Leaders, and in late 2016 published *The Chief Capability Officer.*

He then got to work researching and writing his next book. He published *The Outcome Generation* in late 2018. He spent time talking with people across the world about customer outcomes and outcome programs. It led to his observations that everyone was reinventing the wheel. Everyone was designing their own outcome program. There wasn't a framework companies could use as a starting point.

So, he decided to see if he could develop one. After more than a year of research, interviewing executives and writing, this book emerged.

His sincere hope is that the book will help B2B companies adopt, or improve, an outcome program and thrive as a result.

ACKNOWLEDGEMENTS

The book started with an objective: create a framework that anyone wanting to adopt or improve an outcome program might use. I had some concepts drawn from my own experience running an outcome program. And from all the people who helped me with *The Outcome Generation.*

For this book, close to 150 executives helped me develop and refine the concepts for a framework. Some of the early concepts survived. But many of the ideas in the book came from the people who so generously gave their time.

The bulk of people gave their feedback through one-on-one interviews with me. I have over 250 hours of recordings of these interviews. I learned an enormous amount through this process.

Other people, particularly in the early stages, participated in small group meetings. Again, their input proved invaluable.

And others contributed even more time by also reviewing the working draft of the book. And helping me refine and improve it through their observations and input.

With a process like this, my greatest fear now is that I'll leave someone out who deserves to be recognised.

With that said, let me recognise and thank the people who helped so much in making this book possible (alphabetical order):

Melinda Arredondo, Greg Alexander, Frank Auger, Bruce Ballengee, Richard Bamford, Allison Bantz, Richard Barnett, Sarah Bierenbaum, Nic Bishara, Dave Blake, Darren Blank, Sarah Blocker, Korneel Bouman, Andrew Brock, Graeme Burt, Paul Burton, Patrick Campbell, David Caradonna, Ben Carey, Evan Carlson, Jessica Carroll, David Chia, Anton Chilton, Randy Clapp, Leah Clark, Barry Cochrane, Gereint Collier, Kristina Couchon, Julien Crochet, Prithwi Dasgupta, Stefan de Haar, Kerry Desberg, Verity Dods, Jennifer Dole, Paul Duffy, Dave Duke, Colleen Dwyer, Maranda Dziekonski, David Epperly, Enrique Espinosa,

Kristi Faltorusso, Natalie Fedie, Marie Fenner, Frank Feustel, Margaret Foley, Nidal Fouani, Tim Freeze, Scott Gainey, Brenna Garratt, Jim Goldfinger, Carlos Gonzalez, Tanya Graham, Roz Gregory, Brian Harmon, Andrew Harris, Amelia Hayson, Leigh Hirnisey, Wayne Johnston, Steve Jones, Ramesh Kaza, Kay Kienast, Michael Kirschner, Jeff Klaumann, Arthur Kolayan, Unni Kongot, Duane Kotsen, Alexis La Caze, Aurel Laor, Bob Lelito, Andrew Licence, Kelly Long, Tanya Lord, Joseph Loria, Nick Maley, Luigi Mallardo, JP Mantey, Paul Markovits, Toby Martin, Sam Melrose, Danny Mesrop, Dan Miller, Veronique Montreuil, Andrew Morris, Emma Mourad, Michael Muhlfelder, Matt Myszkowski, Rajie Naicker, Dilip Nair, Rebecca Nerad, Anti Nevalainen, Fai-Keung Ng, Jarred Orr, Victoria Orr, Tom Oryl, Nuno Paiva das Neves, Sal Patalano, Dave Patnaik, Greg Peach, Ziv Peled, Jamie Phillips, Frank Piotraschke, Matt Preschern, Ben Price, Daniel Reese, James Robertson, Jean-Baptiste Roux, Andrew Rowlands, Ted Ruscitti, Neil Ryland, Michael Sachs, Levi Sales, Mike Sasaki, David Satterwhite, Axel Schneider, Indranil Sen, Carmel Segal, Erik Sevilla, Rohan Sheth, Yvonne Sims, Sanders Slaven, Niclas Staberg, John Steinert, Danielle Stewart, Tanya Strauss, Nidhi Suresh, James Taylor, Liza Taylor, Catherine Thomas, Christelle Tissot, Brendan Tremble, Jon Trigg, Matt Trustrum, Bryan Tutor, Colleen Vallinottti, Jussi Vasama, Patrick Vasquez, Revathi Venkatraman, Chris Walz, Kyle Weaver, Min Wei, Mike Weir, Daniel West, Peter Wilson, JP Wirta, Anna Woods, Josh Zamora.

The people who devoted the extra time needed to review the working draft and show me how it could be improved deserve a special thanks. This was a big commitment of time: Bruce Ballengee, Ben Carey, Dave Duke, Andrew Harris, Duane Kotsen, Tanya Graham, Nick Maley, Mike Muhlfelder, Michael Price and Stim Robinson. Thank you so much to this group.

My thanks also to the people who provided the mini case studies. For some, it was an internal battle. For each of you, thank you for the extra contribution you made to the book: Ben Carey, Patrick Campbell, David Caradonna, Natalie Fedie, Ramesh Kaza, Michael Kirschner and Toby Martin.

Matt Trustrum and Liza Taylor acted as business advisers over the past few years. Both helped with the business and with concepts for the book. And Liza gets credit for the term Income from Outcomes. As soon as she coined the term, I knew I had the title for the book.

My thanks to the editors for the book: Gail Barnsley and Damian Clarke. And Anita Saunders for the proofreading. They make a big difference.

Dan Gitau did an amazing job with research, uncovering most of the references used throughout the book.

And finally, special thanks to my assistant Jori Chen, without whose assistance this book would not have happened.

BIBLIOGRAPHY

Adamson, B., Dixon, M., Spenner, P., Toman, N. (2015). *The Challenger Customer.* Portfolio Penguin.

Akhter, N. (2018). 'Outcome based Service Delivery Model'. Medium. Retrieved from https://medium.com/@nadeemakhte/outcome-based-service-delivery-model-d8da7c4676af

Amirault, P. et al. (2017). 'It's Time to End Quarterly Reporting'. Nortonrosefulbright.com. https://www.nortonrosefulbright.com/en/knowledge/publications/850a7ca1/its-time-to-end-quarterly-reporting

Babar, Z. (2015). 'Pharmaceutical policies in Australia and New Zealand'. *Journal Of Pharmaceutical Policy And Practice*, *8*(S1). https://doi.org/10.1186/2052-3211-8-s1-k5

Baines, T. (2020). 'Comment: How servitization can give manufacturers a competitive edge'. The Engineer. Retrieved from https://www.theengineer.co.uk/comment-how-servitization-can-give-manufacturers-a-competitive-edge/

Balaji, S. 'SaaS Pricing 101 – Scale Value, Not Usage (It's Harder than You Think)'. ChaaS Dispatch. Retrieved from https://www.chargebee.com/blog/saas-pricing-scale-value-not-usage/

Boston Consulting Group, Konikoff, J, Mayer, A. Florian Schmieg, A.F. Stephan, M. and Trifonov, S. February 23, 2021. 'How Delivering Outcomes Changes Everything'. https://www.bcg.com/publications/2021/value-of-outcome-based-business-models

Catapult Energy Systems. (2020). 'Baxi Heating Trial Heat-as-a-Service Bundle – Energy Systems Catapult'. Energy Systems Catapult. https://es.catapult.org.uk/case-studies/baxi-heating/

Covey, S.R. (1988). *The Seven Habits of Highly Effective People.* Free Press.

Dove, M. 'Outcome Selling Is the Only Way Forward'. Technology Services Industry Association. https://www.tsia.com/blog/outcome-selling-is-the-only-way-forward

Godin, Seth: 'Seth's Blog'. February 3, 2019. https://seths.blog/2019/02/getting-the-word-out-2/

Gill, J., Kwan, A., Shrouek, M. (2021). 'Scaling Xaas: Outcome-based Monetization Models'. Deloitte Insights. https://www2.deloitte.com/xe/en/insights/focus/industry-4-0/xaas-outcome-based-pricing.html

IEDP (2017). 'Servitization A Model for Growth'. https://www.iedp.com/articles/servitization-a-model-for-growth/

James, N., Lawson, K., & Acharya, Y. (2020). Evidence on result-based financing in maternal and child health in low- and middle-income countries: a systematic review. *Global Health Research And Policy*, *5*(1). https://ghrp.biomedcentral.com/articles/10.1186/s41256-020-00158-z

Jee, M. & Or, Z. (1999). 'Health Outcomes in OECD Countries'. OECD Labour Market and Social Policy Occasional Papers. https://doi.org/10.1787/513803511413 (Note: now outdated, but has information about the 1997 OECD meeting)

Lauterbach, K. (2016). 'Germany's Model for Drug Price Regulation Could Work in the US'. Healthaffairs.org. Retrieved 11 March 2021, from https://www.healthaffairs.org/do/10.1377/hblog20161229.058150/full/

Lecinski, J. (2011). 'Winning the Zero Moment of Truth'. https://www.thinkwithgoogle.com/future-of-marketing/emerging-technology/2011-winning-zmot-ebook/

Lerner, J., Ye Li, Valdesolo, P. and Kassam, K. 'Emotion and Decision Making'. *Annual Review of Psychology.* https://www.annualreviews.org/doi/pdf/10.1146/annurev-psych-010213-115043

LinkedIn (2020). 'LinkedIn Age-of-Agility Global Report'. https://business.linkedin.com/content/dam/me/business/en-us/marketing-solutions/products/pdfs/LinkedIn_Age-of-Agility_Global_Report.pdf

Mjåset, C., Ikram, U., N. S. Nagra, & T. W. Feeley. (2020). 'Value-Based Health Care in Four Different Health Care Systems'. Catalyst.nejm.org. Retrieved from https://catalyst.nejm.org/doi/full/10.1056/CAT.20.0530

Moore, G. (October 21, 2021).' The Ongoing Evolution of Software-as-a-Service: Where Next?' https://www.linkedin.com/pulse/ongoing-evolution-software-as-a-service-where-next-geoffrey-moore/?trackingId=7hiBsI%2BwTVm3k%2BAg%2By7Oow%3D%3D

Morfin, Z. 'Short-Termism Hinders Innovation; How To Mitigate It – Corporate Board Member'. https://boardmember.com/short-termism-hinders-innovation/

Nebuloni, G., Hernandez, D., & Carter, P. (2018). 'IDC Servitization Barometer: Charting Your Path to New Revenue Streams'. Retrieved from https://www.ifs.com/sitecore/media-library/assets/2020/04/23/idc-servitization-barometer-charting-your-path-to-new-revenue-streams/

Owen, A., Spinks, J., Meehan, A., Robb, T., Hardy, M., & Kwasha, D. et al. (2008). 'A new model to evaluate the long-term cost effectiveness of orphan and highly specialised drugs following listing on the Australian Pharmaceutical Benefits Scheme: the Bosentan Patient Registry'. *Journal Of Medical Economics, 11*(2), 235-243. https://doi.org/10.3111/13696990802034525

PA Consulting Group. *From Products to Services* [EBook]. http://www2.paconsulting.com/rs/526-HZE-833/images/PA%20Servitization%20Report.pdf

Profitwell; Campbell, P, September 11, 2018. 'Outcome Based Value Metrics Reduce Churn, Increase Revenue'. https://www.profitwell.com/recur/all/outcome-based-value-metrics- for-growth

Ramadan, A., Petersen, D., Lochhead, C., with Maney, K. *Play Bigger.* 2016. HarperCollins e-books.

Serafeim, G. *Harvard Business School Publications.* January-February 2019. Corporate Purpose and Financial Performance https://www.hbs.edu/faculty/Pages/item.aspx?num=54523

Smith, R. 'The Founder of Qualtrics on Reinventing an Already Successful Business'. *Harvard Business Review* magazine. July 2021. https://hbr.org/2021/07/the-founder-of-qualtrics-on-reinventing-an-already-successful-business

Salesforce. (2018). *State of the Connected Customer 2nd Edition.* Salesforce.com. https://www.salesforce.com/content/dam/web/en_us/www/documents/e-books/state-of-the-connected-customer-report-second-edition2018.pdf

Starita, L., September 2019. 'The Secret to B2B Account Growth'. Gartner. https://www.gartner.com/en/marketing/insights/articles/the-secret-to-b2b- account-growth

NOTES

(1) In researching the book, I asked executives about the late engagement of vendors in the buying cycle. Each executive interviewed had a similar comment. They said that every vendor knows about the late engagement, but few have tackled the problem well. There is lots of other research supporting this perspective, including from the authors of *The Challenger Customer* (Adamson, B., et al., 2015).

(2) *The Challenger Customer* authors also did a great job describing the difficulty of getting alignment of focus in a customer or prospect.

(3) For me, the concept of a Product Outcome and a Success Outcome began with the quote, 'people-don't-want-to-buy-a-quarter-inch-drill-they-want-a-quarter-inch-hole'. The expression has an interesting history. It's most often associated with Harvard professor Theodore Levitt. In 1969 he published *The Marketing Mode: Pathways To Corporate Growth*, and he attributed the saying to Leo McGivena: 'Last year 1 million quarter-inch drills were sold,' Leo McGivena once said, 'not because people wanted quarter-inch drills but because they wanted quarter-inch holes.' However, references to the concept have been found as far back as 1942 (see https://quoteinvestigator.com/2019/03/23/drill/).

Regardless, the concept of the Product Outcome is based on the drill bit and the hole in the wall. The question we really need to answer, however, is why the person wants a hole in their wall. The answer is the Success Outcome, which was introduced in *The Outcome Generation.*

(4) I first heard the concept of the travel poster on an excellent course called Key Person of Influence. Dent Global run this course. You can learn more at https://www.dent.global/

(5) The chapter on pricing models is based on input from several sources. But the most influential is the work by Patrick Campbell at Profitwell. You'll find more at https://www.profitwell.com/

(6) This change management guide is based on work pioneered by John Kotter. You'll find more on his work at: https://www.kotterinc.com/8-steps-process-for-leading-change/

TABLE OF FIGURES

INDEX OF KEY TERMS